PERGAMON INTERNATIONAL LIBRARY
of Science, Technology, Engineering and Social Studies
*The 1000-volume original paperback library in aid of education,
industrial training and the enjoyment of leisure*
Publisher: Robert Maxwell, M.C.

Applied Geophysics for Geologists and Engineers

THE ELEMENTS OF GEOPHYSICAL PROSPECTING

THE PERGAMON TEXTBOOK
INSPECTION COPY SERVICE

Related Pergamon Titles of Interest

Books

ANDERSON & OWEN:
The Structure of the British Isles, 2nd Edition

ANDERSON:
The Structure of Western Europe

BATES *et al:*
Geophysics in the Affairs of Man

CONDIE:
Plate Tectonics and Crustal Evolution, 2nd Edition

KELLER:
Electrical Methods in Geophysical Prospecting

OWEN:
The Geological Evolution of the British Isles

PATERSON:
The Physics of Glaciers, 2nd Edition

ROBERTS:
Introduction to Geological Maps and Structures

SIMPSON:
Geological Maps

*WHITELEY:
Geophysical Case Study of the Woodlawn Orebody, NSW, Australia

**Journals*

Computers and Geosciences
Continental Shelf Research
Journal of African Earth Sciences
Journal of Structural Geology

*Not available under the terms of the Pergamon inspection copy service.
**Free specimen copy of any Pergamon journal available on request.

Applied Geophysics for Geologists and Engineers
The Elements of Geophysical Prospecting

by

D. H. GRIFFITHS
Professor of Geophysics,
University of Birmingham

and

R. F. KING
Reader in Geophysics,
University of Birmingham

(A Second Edition of
Applied Geophysics for Engineers and Geologists)

PERGAMON PRESS
OXFORD · NEW YORK · TORONTO · SYDNEY · PARIS · FRANKFURT

U.K.	Pergamon Press Ltd., Headington Hill Hall, Oxford OX3 0BW, England
U.S.A.	Pergamon Press Inc., Maxwell House, Fairview Park, Elmsford, New York 10523, U.S.A.
CANADA	Pergamon Press Canada Ltd., Suite 104, 150 Consumers Road, Willowdale, Ontario M2J 1P9, Canada
AUSTRALIA	Pergamon Press (Aust.) Pty. Ltd., P.O. Box 544, Potts Point, N.S.W. 2011, Australia
FRANCE	Pergamon Press SARL, 24 rue des Ecoles, 75240 Paris, Cedex 05, France
FEDERAL REPUBLIC OF GERMANY	Pergamon Press GmbH, Hammerweg 6, D-6242 Kronberg-Taunus, Federal Republic of Germany

First edition 1965
Reprinted 1974, 1975, 1976
Second edition 1981
Reprinted with corrections 1983

British Library Cataloguing in Publication Data

Griffiths, Donald Harrison
Applied geophysics for geologists and
engineers. -2nd ed.
1. Prospecting—Geophysical methods
I. Title   II. King, Roy Farell
III. Applied geophysics for engineers
and geologists
622'.15   TN269   80-40474
ISBN 0-08-022071-1 (Hardcover)
ISBN 0-08-022072-X (Flexicover)

Printed in Great Britain by A. Wheaton & Co. Ltd., Exeter

Preface to the Second Edition

AN elementary textbook is much concerned with the fundamentals of a subject and these do not change with time. In this respect the second edition of our book has been little modified. In every other it is much altered for the subject has advanced greatly since we gathered together the material for the original version. The changes have been brought about mainly by developments in electronics and in the science of computing. The former have led to the appearance of much lighter, more sensitive and altogether much improved instruments, making it possible to collect better data far more quickly. The latter have made possible the rapid reduction of the very large amounts of data that can now be collected during a survey but even more important they make possible the use of far more sophisticated interpretation techniques than formerly. It is no exaggeration, therefore, to say that applied geophysics has moved into a new era in the interim since this book was first conceived.

Thus, though the sections in the first edition dealing with basic ideas remain little modified, extensive rewriting has been necessary and certain alterations have had to be made to the general plan to bring about the required changes of emphasis.

As the preface to the first edition states, the book was originally written with the needs of the civil engineer in mind, though the interests of others were not forgotten. The book has, however, proved to be popular with students of geology and the alterations to the balance of the material that have been made have taken this into account. For example, more attention has been paid to applications of geophysics in the search for ore minerals, this being reflected in considerable strengthening of the sections dealing with electromagnetic and induced polarization techniques. However, the bias of the book remains, as in the first edition, towards the seismic and resistivity methods which dominate the field of engineering geophysics, and it does not pretend to be a comprehensive textbook giving weight to techniques in proportion to their usage in the industry as a whole.

Our revision of the book has been stimulated and assisted by the comments of our colleagues, in particular by Dr W. H. Owens, Dr K. R. Nunn and Dr R. D. Barker, and we gratefully acknowledge their help. Mrs B. M. Holland has again prepared the manuscript for publication.

D. H. GRIFFITHS
R. F. KING

Preface to the First Edition

THE science of geophysics was originally concerned with the study of the shape and structure of the earth as a whole, but from about 1920 onwards geophysical techniques have been increasingly used on a smaller scale in the search for minerals, and in particular for petroleum, in the upper few miles of the earth's crust. Strangely enough it is only much more recently that geophysical methods have been seriously applied to the problem of shallow geological structure, in which depths are measured in tens rather than hundreds or thousands of feet, and civil engineers have therefore hitherto made little use of them. Now, however, engineers recognize geophysics as a tool which can often give important information about a site as effectively and more cheaply than a very large number of bore-holes, although some boreholes are always necessary before a geophysical survey can be uniquely interpreted. This means that the average civil engineer, although he will probably never have to assume full responsibility for a geophysical survey, must know how such a survey can (or cannot) be expected to help him in given circumstances, and must be in a position to discuss its results intelligently with the contractors who carry it out.

This book is written primarily with the needs of civil engineers in mind, but it is our experience that geologists, mining engineers and petroleum engineers also need to understand geophysical techniques even though they may never specialize as geophysicists. The scope of the book has therefore been broadened to include some discussion of geophysical methods used in prospecting for oil and other minerals, even though these methods will be of little or no interest to the civil engineer. However, it has no pretensions to completeness, even as an introduction to geophysics, and must be regarded rather as a handbook for the student whose main interests lie elsewhere, but who realizes that geophysical methods can provide important information on the basis of which he is likely to have to make decisions.

Our acknowledgements are due to several of our colleagues for helpful discussion and criticism of the manuscript, and to Mrs W. Holland for typing it.

D. H. GRIFFITHS
R. F. KING

Contents

CHAPTER 1

Introduction; Units

1.1 Introduction

THE objects of a geophysical survey are to locate subsurface geological structures or bodies and where possible to measure their dimensions and relevant physical properties. In oil prospecting structural information is sought because of the association of oil with particular features such as anticlines in sedimentary rocks. In mining geophysics the emphasis is on detection and determination of physical properties. Though mineral ore bodies give distinctive and measurable geophysical indications they are often of irregular shape and occur in rocks of complex structure, making precise quantitative interpretation difficult or impossible. In site investigation engineers may be interested in both structure and physical properties. Variations in bedrock depth are often needed on major construction sites and the mechanical properties of the overburden may be important when heavy loads have to be sustained.

A geophysical survey consists of a set of measurements, usually collected to a systematic pattern over the earth's surface by land, sea or air, or vertically in a borehole. The measurements may be of spatial variations of static fields of force – the gradients of electrical, gravitational or magnetic "potential" – or of characteristics of wave fields, more particularly of travel-times of elastic (seismic) waves and amplitude and phase distortions of electromagnetic waves. These force and wave fields are affected by the physical properties and structure of the subsurface rocks. Since physical properties are determined to a considerable degree by lithology, discontinuities in physical properties often correspond to geological boundaries and thus any structural problem becomes one of interpreting the fields at the surface in terms of these discontinuities. The ease with which this can be done depends on many factors, of which the complexity of the structure and the degree of contrast in physical properties of the rocks forming it are particularly important. Clearly, in choosing the geophysical technique to be used to study a problem the contrasting properties of the subsurface rocks and their homogeneity within a particular formation are important factors to be considered.

The properties of rocks of which most use is made in geophysical prospecting are elasticity, electrical conductivity, density, magnetic susceptibility and remanence and electrical polarizability. To a lesser extent other properties such as degree of radioactivity are also utilized.

All matter has a gravitational effect and lateral changes in density within the earth therefore produce small but often measurable variations in gravity over the surface. Likewise many rocks contain small quantities of magnetic minerals and therefore show a degree of magnetization. Differences in intensity of magnetization between rocks arising from differences in magnetic susceptibility or permanent magnetization give rise to spatial variation in the resultant magnetic field, again measurable over the earth's surface. From the form of the surface gravitational or magnetic fields deductions about subsurface structure may be made; though because of the inherent ambiguity of these "potential field" methods, geological or other geophysical information is needed as a constraint if useful solutions are to be obtained.

Gravitational and magnetic surveys make use of natural fields of force. Most seismic and electrical (including electromagnetic) methods, which involve the elastic and electrical properties of rocks, necessitate introducing energy into the ground. Since the source is under control the source to detector distance can be varied. This makes it possible to interpret the results with much less ambiguity than is possible when gravity and magnetic fields are being utilized.

In some electrical methods either direct or very low frequency current is passed into the ground through electrodes. The form of the electrical field at the surface is determined by the disposition of the electrodes and the distribution of electrical conductivity in the subsurface. It is this surface field that is measured. Where direct or low frequency alternating current is used and applied directly to the ground the measurement is essentially one of the potential gradient (i.e. voltage difference between two measuring electrodes). It may however be presented by the instrument used as a resistance. When direct current is passed into the ground some metallic ores exhibit electrical polarization. They become electrically charged and when the current is switched off a transient discharge lasting a few seconds is noted. The Induced Polarization method makes use of this effect. It is also practicable to inject energy into the ground by induction using a coil carrying an alternating current with a frequency of a thousand or so cycles per second. There is however no direct electrical contact with the ground. In these so-called electromagnetic methods the alternating magnetic field produced by the transmitting coil induces eddy currents in good conductors in the ground, manifest at the surface as secondary fields and so measurable by means of a search coil.

There are two distinct seismic techniques, one making use of the reflection and the other of the refraction of elastic waves in rocks. The reflection method which is universally employed in the search for oil is by far the most highly developed of all prospecting methods and where applicable yields the most information. Because rocks have differing densities and elastic properties elastic (i.e. sound) waves are propagated through them with different velocities and are reflected and refracted at interfaces across which there is a change of properties. Thus in a sedimentary layered sequence an elastic wave generated by an explosion at the surface is in part reflected and refracted back to the surface by the various interfaces, and from a knowledge of the travel time of the wave to a number of points on the surface the form and dimensions of the subsurface structure can be calculated.

Many factors, geological, economic, logistic and what we might call geophysical govern the choice of method for a particular survey. There is no alternative to reflection seismology for the mapping of structure in sedimentary rocks at depths of many thousands of feet. The method has been developed to the point at which it is possible to build up an accurate and very full picture of the subsurface, even when this shows considerable complexity. For small scale problems such as are met in civil engineering or hydrogeological work the choice is between seismic refraction and electrical resistivity.

In mining geophysics, where metallic ores are being sought, the geology is usually too complex for seismic techniques to be used. Here the marked physical contrast between the ores and the host rocks makes electromagnetic, magnetic and sometimes even gravitational methods practicable. Interpretation is at best only semi-quantitative, for such bodies tend to be irregular in shape and inhomogeneous. Nevertheless useful indications of size, depth and quality can be obtained.

In many instances more than one method will be used to survey the same ground. The search for oil may start with gravity and airborne magnetic work as a preliminary to seismic shooting in localities determined by interpretation of the earlier regional surveys. Combining electromagnetic, magnetic and gravity data may make it possible to decide whether certain indications are of valuable metallic ores or merely of concentrations of uneconomic minerals.

Having chosen the method or methods and the appropriate instruments the layout of the survey has to be decided. This is very much a matter of station or line pattern and density. How much data will be required to obtain an adequate interpretation? This is determined to some extent by the economics of the situation. For example, survey lines must pass over or at least be close to shallow ore bodies if these are to be detected. The minimum separation of lines to be used is obviously very much influenced by the size and value of the ore, for the probability of finding an ore body worth a certain sum increases as line spacing is decreased, thus putting up survey costs. The more fundamental problem of "noise" also has to be considered. Apart from purely instrumental limitations every measurement is subject to the effect of random, local variations in subsurface properties. The greater this "noise" compared with the magnitude of the disturbance or "anomaly" due to the body being sought, the more closely spaced the data need to be if the anomaly is to be picked out with sufficient accuracy. It is thus background rather than instrumental limitations that controls detectability. It is always present to some degree and for this reason interpretations based on single or widely spaced traverses can be very suspect. Inhomogeneity may in some instances become so serious that unambiguous interpretation becomes impossible. At first sight it might seem a simple problem to find the depth to a relatively flat bedrock beneath sandy overburden, apparently an example of a single horizontal interface between two media of widely differing electrical conductivity. Yet this might prove impossible because of the inhomogeneous nature of the overburden, for variations in clay content, grain size and water content produce marked changes in conductivity. The problem thus ceases to be one in which there is only one interface; there are now many, and of these some will be laterally impersistent. Thus interpretation based on the simple assumption of a single interface is invalid and the complexity may well be such that no solution is possible. Of course no overburden is ever truly uniform; all interpretation techniques allow of some latitude in this respect and will work if the departures from homogeneity are not too great. The difficulty is to recognize, preferably in advance, that what appears simple is really complex. It also has to be remembered that in many areas of geophysics even when using the most advanced methods satisfactory interpretations can only be achieved when the structures are relatively simple.

The final decision to be taken in prospecting and site investigation is whether or when to drill. In civil engineering where the depths of investigation are small and high accuracy is required it may pay to dispense with geophysics and drill from the outset. As depths or distances to be covered increase, particularly if the geology is simple, geophysics will be increasingly used, controlled by widely spaced but adequate boreholes. In prospecting for oil the structures to be discovered lie at great depth, making exploratory drilling on a large scale prior to geophysical survey out of the question. In any case the techniques of reflection seismology are so advanced that this is quite unnecessary.

Generally, both economic and scientific factors have to be considered in deciding a drilling programme. The cost of drilling to map accurately the undulating surface of the bedrock on a site may be very high relative to the cost of using geophysics but if the economic penalties of inaccuracy are substantial it may be cheaper to drill. On the other hand, where a large area must be covered in detail but high accuracy is not essential, geophysics can be the obvious answer.

1.2 Units in Geophysics

The SI (Système International) units used in this book have come into common use since the first edition was published, but older units are still to be found, and a note on conversions may be of some value.

Electrical units

These are the least affected by the introduction of SI, which uses the same units (volt, ampere, ohm etc.) as the "practical" version of the e.m.u. system. The only real change is that resistivity is now measured in ohm-metres (Ωm) rather than in ohm-cm or ohm-ft.

$$1 \text{ ohm-cm} = 10^{-2}\,\Omega\text{m}$$
$$1 \text{ ohm-ft} \approx 0.3\,\Omega\text{m}$$

Use of this latter conversion factor (more precisely 0.305) rather than its reciprocal in all conversions involving feet and metres makes for easy mental conversions to better than 2 per cent accuracy.

Conductivity is of course measured in mho m^{-1}. The "reciprocal ohm" or mho is now also known as the Siemens (S), so that conductivities may be expressed in S m^{-1}.

Seismic velocity

The SI unit is the metre per second (m s^{-1}) but seismologists commonly use the kilometre per second (km s^{-1}), a unit which is conveniently also thought of as the metre per millisecond (m ms^{-1}) in applied seismology.

Gravity

If a gravitational field is conceived as an acceleration, its SI unit will be the m s^{-2}; if as a force per unit mass it will be measured in Newtons per kilogram (N kg^{-1}), the same unit by a different name. The "gravity unit" is the μm s^{-2} (or μN kg^{-1}) so that

$$1 \text{ g.u.} \equiv 10^{-6}\text{m s}^{-2} \equiv 10^{-6}\text{N kg}^{-1} \equiv 10^{-1}\text{mGal}$$

since the old "milligal" (mGal) unit was defined as

$$10^{-3}\text{cm s}^{-2} = 10^{-5}\text{m s}^{-2}.$$

The SI unit of density is the kg m^{-3}: the density of water is 1000 kg m^{-3}, and to avoid the use of large numbers it is convenient to express densities in tonnes per cubic metre (t m^{-3}) so that they have the same numerical value as in the old units of g cm^{-3}, since 1 tonne = 1000 kg.

Magnetism

There has been some confusion in the application of the SI to magnetic units, particularly in geophysics, since at least three slightly differing conventions have been in use. All these conventions are based on "rationalized" units, in which the force between two magnetic poles m_1, m_2 *in vacuo* becomes $m_1 m_2 / 4\pi\mu_o r^2$ rather than $m_1 m_2 / \mu_o r^2$. The constant μ_o, known as the permeability of free space, has the value $4\pi \times 10^{-7}$ in SI units and unity in the old c.g.s. electromagnetic units. The factor of 4π introduced by rationalization means that although magnetic susceptibility is a dimensionless ratio its value in SI units will be a factor of 4π larger than that appropriate to the (unrationalized) c.g.s. e.m.u. system.

In the particular (Sommerfeld) convention used in this book, magnetizing field F and intensity of magnetization J are both measured in amperes per metre ($A\ m^{-1}$) and are related to the magnetic induction B in a magnetized medium, which is measured in teslas (T), through $B = \mu_o(F + J)$.

In the treatment of this book, the geomagnetic "field" and its anomalies are taken to be fields of the induction B rather than of F; thus outside magnetic material the normal geomagnetic field $B_g = \mu_o F_g$. The intensity of magnetization J produced in a body of susceptibility k will be kF_g, but we work only with the quantity $\mu_o J = k\mu_o F_g = kB_g$ and calculate from it the "B" field produced by a magnetized body. Magnetic anomaly fields are then conveniently measured in nanoteslas (nT), a unit which is the same as the gamma (γ) used in the c.g.s. system, since $1\gamma = 10^{-5}$ gauss and 1 tesla $= 10^4$ gauss, the gauss being the unit of magnetic induction in the e.m.u. system.

CHAPTER 2

The Seismic Method

2.1 Outline of the Method

IN the seismic method an elastic pulse or a more extended elastic vibration
is generated at shallow depth, and the resulting motion of the ground at
nearby points on the surface is detected by small seismometers or
"geophones". Measurements of the travel-time of the pulse to geophones
at various distances give the velocity of propagation of the pulse in the
ground. The ground is generally not homogeneous in its elastic properties,
and this velocity will therefore vary both with depth and laterally. Where
the structure of the ground is simple the values of elastic wave velocity
and the positions of boundaries between regions of differing velocity can
be calculated from the measured time intervals. "Velocity" boundaries
usually coincide with geological boundaries and a cross-section on which
velocity interfaces are plotted may therefore resemble the geological
cross-section, although the two are not necessarily the same. The method
has been applied most extensively in the search for oil, or rather for the
geological structures most favourable for its accumulation, at depths
generally of the order of a few kilometres. Direct detection of oil is rarely,
if ever, possible, but some success has now been attained in the location
of gas accumulations which considerably reduce the seismic velocity of
the sedimentary rocks in which they occur.

Seismic methods are also of major importance in the fields of engineering site investigation and hydrogeology. The depths of interest here lie in
the range of tens of metres to no more than a few hundred metres, and
the problems which may be solved range from the estimation of the depth
of high-velocity "bedrock" or of a well defined water table to the
evaluation of the mechanical and hydrological properties (degree of
fracturing, porosity, degree of saturation etc.) of a concealed foundation
material or aquifer.

8

In prospecting for minerals other than petroleum seismic methods are little used, mainly because such mineral deposits are often small in extent with similar seismic velocities to their host rocks, and tend to occur in geologically complex situations which lead to interpretational difficulties. In any application of seismology to small-scale problems it is important to bear in mind that the resolution of the method, that is, its ability to separate closely-spaced structures, is limited to dimensions not much less than the length of the seismic waves, found by dividing their velocity by their frequency. Seismic velocities in soils and crustal rocks range between about 200 and 6000 m s^{-1} (see §2.4), and the frequencies used in seismic prospecting range from about 10 to 200 Hz, so that wavelengths from 1 m to 600 m may be encountered, typical wavelengths in petroleum prospecting being about 100 m, and in site investigation perhaps 3–30 m.

2.2 Generation and Propagation of Seismic Waves

The various methods of generating seismic pulses and extended waves for prospecting are discussed in §3.2. However the pulse or wave is generated the motion recorded by a detector (usually the vertical component of ground velocity) is very complex, partly because different types of elastic wave, travelling with different velocities, are generated together, and partly because each of these wave types can, by reflection and refraction at interfaces, follow several different paths to the geophone. The *seismogram* or record of the ground motion produced by a short pulse will therefore extend over a much longer time than the pulse duration.

2.21 *Wave types*

In an elastically homogeneous ground, stressed suddenly at a point *S* near its surface (Fig. 2.1), three elastic pulses travel outwards at different speeds. Two are *body waves*; that is, they are propagated as spherical fronts affected to only a minor extent by the free surface of the ground, and the third is a *surface wave* which is confined to the region near this free surface, its amplitude falling off rapidly with depth in the body of the medium.

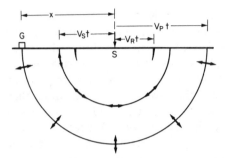

Fig. 2.1   Pulse fronts of the *P, S* and *R* waves at a time *t* ms after their initiation at the point *S*.

The two body waves differ in that the ground motion within the pulse is in the direction of propagation (i.e. radial) in the faster or "*P*" (primary) wave, but normal to it (i.e. tangential to the pulse front) in the slower "*S*" (secondary) wave. The stresses in the *P* wave, which is a longitudinal wave like a sound wave in air, are thus due to uniaxial compression, whilst during the passage of an *S* wave the medium is subject to shear stress. The velocities V_P and V_S of these two waves are related to the elastic constants and density of the medium by the equations

$$V_P = \sqrt{\left(\left(k + \frac{4}{3}n \right) / \rho \right)} \tag{2.1}$$

$$V_S = \sqrt{(n/\rho)} \tag{2.2}$$

where *k* is the bulk modulus (or "incompressibility"), *n* the shear modulus and ρ the density. The shear wave travels at little more than half the velocity of the *P* wave in most rocks, and is not propagated at all in fluids, for which $n = 0$. The value of V_P/V_S can be shown to depend only on Poisson's ratio σ (see §2.4).

The *surface wave* travels more slowly than either body wave, and is generally complex. In the special case of a homogeneous ground which we are now considering, the surface disturbance is caused entirely by the wave known as a *Rayleigh wave*, in which both vertical and horizontal components of ground motion are present. The horizontal ground motion is of rather smaller amplitude than the vertical, and is 90° out of phase with it, so that the resultant path of an element of the medium during the passage of a Rayleigh wave cycle follows an ellipse lying in the plane of propagation. The ground motion becomes negligibly small within a distance from the free surface of the same order as the wavelength of the disturbance. The velocity V_R of the Rayleigh wave is only about 10 per cent less than the body shear wave velocity, the ratio V_R/V_S again depending only on Poisson's ratio.

If the ground were perfectly elastic, the amplitudes of the waves would decrease with range from their source simply as a result of the spreading of the wave energy (proportional to the square of the amplitude) over the increasing wave-front area. This area will vary as the square of the range for spherically spreading body waves and linearly with range r for cylindrically spreading surface waves. Body and surface waves would thus have amplitudes varying as r^{-1} and $r^{-\frac{1}{2}}$ respectively. The form of the pulse is determined by the relative amplitudes of the different frequency components which make up its *spectrum* which can be determined by Fourier analysis. Since these spectral components are equally affected by geometrical spreading, one would not expect the pulse form to be changed during its propagation. Real earth materials, however, are imperfectly elastic, leading to energy loss and attenuation of the seismic waves, that is, to an amplitude reduction more rapid than would be expected from geometrical spreading alone. This attenuation is more pronounced for less consolidated rocks and is also greater for higher frequencies, selective loss of the high frequency components of a pulse leading to its progressive broadening in time as the pulse propagates.

The three types of wave appear in order on the idealized seismogram of Fig. 2.2, which is a graph of ground motion against time at a particular

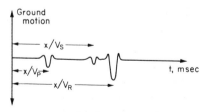

Fig. 2.2   Idealized seismogram of the ground motion at a distance x from a source in homogeneous ground.

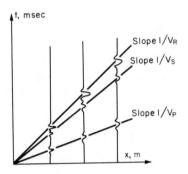

Fig. 2.3   Travel-time graph constructed from a set of seismograms such as that of Fig. 2.2 recorded at various ranges x.

geophone G a distance x from the "shot-point" S. The time zero is, of course, the time of the shot, and it is clear that the three velocities V_P, V_S, V_R could be found from this record. In practice, this determination is made by combining on a *time-distance graph* as in Fig. 2.3 the information from several geophones at various distances from the shot. Each seismogram will give three travel-times for a particular value of x, and the points should lie on three straight lines passing through the origin, though in reality they will be scattered about these lines as a result of inhomogeneities in the ground. It is usual to take the *distance* axis of the graph to be horizontal, and so the slopes of the graphs $t=x/V_P$, $t=x/V_S$, $t=x/V_R$ measured from this axis will be equal to $1/V_P$, $1/V_S$ and $1/V_R$. The velocities themselves can be quickly found by reading off from the graph the distance travelled in, say, 10, 50 or 100 ms. If distances are measured in metres and times in milliseconds (travel times can be readily measured to $\pm$ 1 ms or better) it is convenient to measure seismic velocities in metres per millisecond (m ms^{-1}): this is the same unit as the kilometre per second (km s^{-1}) commonly used in crustal and global seismology. The range of P-wave velocities encountered in applied seismology becomes 0.2–6 m ms^{-1} in these units (see §2.4).

   If the seismograms from a set of geophones are arranged side by side with their time axes vertical, and separated by horizontal distances proportional to the actual geophone separation on the ground, the travel-time graph can be traced directly through the various pulses apparent on the seismograms. This has been done in Fig. 2.4 for a set of real seismograms obtained on ice, which is one of the better natural approximations to a homogeneous solid medium. The slopes corresponding to the velocities of the three main wave types are clearly shown. A later event corresponding to a reflection from the base of the ice (see §2.2) can also be seen.

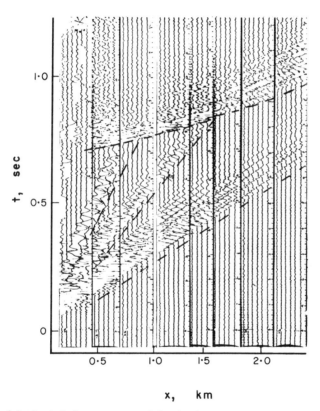

x,   km

Fig. 2.4   A set of seismograms recorded on ice (from Holtzscherer and Robin, 1954). The broken lines have been drawn through the main events recorded, giving approximate travel-time graphs for the *P*, *S* and surface waves. The curved line is drawn through an event which can be identified as the *P*-wave reflected from the bedrock below the ice (see Fig. 2.9).

In geophysical prospecting, as distinct from large-scale earthquake seismology, shear waves have not so far been greatly used. The stress system near a shot is predominantly one of radial compression, and therefore preferentially generates a compressional pulse. When this is refracted and reflected at discontinuities in the ground some transformation into shear pulses occurs, but conventional geophones are sensitive only to the vertical component of ground motion, and this often results in further discrimination against shear motion, which is horizontal for pulses returned at steep angles to the surface. It is therefore generally assumed in the first place that the events on a seismogram are all either P waves or surface waves, although the possibility of the appearance of S-wave events cannot be entirely neglected. When the impact of a weight or hammer is used as a source, S waves may be of much greater importance, and in some cases it may be desirable to deliberately generate S waves and to record them using horizontal component geophones so that V_S can be determined as well as V_P. In the following simplified treatment, only P waves will be considered.

2.22 *Ray paths in a layered medium*

A diagram like Fig. 2.1 which shows the position of the pulse front at a single moment of time tells us nothing about its progress over the rest of the path between shot-point and geophone. In seismology it is the travel-time over this path that is of principal interest, and so it is usually more convenient to represent the path by means of rays. A ray is a line which is drawn so as to be always perpendicular to successive portions of the pulse front (this is strictly true only for isotropic materials), and is therefore a path along which the energy of the pulse is propagated. To any one pulse front there corresponds an infinite number of rays, but only a selected few of these are drawn on any one diagram, which therefore gives an incomplete picture. More information is given by a pulse-front diagram on which the positions of the front at fixed intervals of a few milliseconds are drawn. We shall use either ray or pulse-front diagrams as seems most appropriate, sometimes combining the two on one figure. The rays can always be distinguished as heavier lines with arrowheads.

One horizontal interface. The real earth, which in fact often consists of stratified material, is usually best approximated by a layered medium, each layer having a constant velocity or one changing in a simple and regular way with depth. The interfaces between the layers may be inclined at any angle to the horizontal and to each other, but the model can be treated most simply when the layering is horizontal. We shall first consider the case of one horizontal interface at a depth h between media in which the P wave velocities are V_1 and V_2, V_2 being the greater. Figures 2.5 and 2.6 are ray diagrams showing three possible paths for body waves between a shot S and a geophone G: we shall consider them in the order in which the rays are numbered.

The first path (Fig. 2.5) is of course the same as the path of the surface wave, but we are now discussing the P wave only, and this direct wave

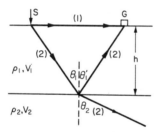

Fig. 2.5   The paths of a direct ray (1) and of a ray (2) of the same pulse which is reflected and refracted at an interface.

travels with the velocity V_P, which is much greater than the Rayleigh wave velocity V_R.

Any wave meeting an interface between media of differing velocities is partly reflected and partly refracted. The laws of reflection and refraction are the same as those familiar in optics, so that for ray 2 in Fig. 2.5, $\theta_1 = \theta_1'$ and

$$\sin \theta_1 / V_1 = \sin \theta_2 / V_2 \quad \text{(Snell's Law)} \qquad (2.3)$$

The "refractive index" of the interface is thus $\sin \theta_1 / \sin \theta_2 = V_1 / V_2$ and for $V_2 > V_1$ the refraction is *away* from the normal as shown.

The amplitudes A_r and A_t of the reflected and transmitted waves vary in a complicated way with the angle of incidence, but the reflection coefficient $r = A_r/A_i$ for a normally incident wave of amplitude A_i depends only on the properties of the media and is given by

$$r = \frac{\rho_2 V_2 - \rho_1 V_1}{\rho_2 V_2 + \rho_1 V_1} = \frac{Z_2 - Z_1}{Z_2 + Z_1} \qquad (2.4)$$

so that $-1 \leqslant r \leqslant +1$.

It depends on the "*acoustic impedances*" Z_2 and Z_1 of the media rather than on their velocities alone, but changes in the density ρ are in practice no more than about ± 20 per cent, while seismic velocity may change by ± 50 per cent. It is clear that an interface with a good velocity contrast (strictly speaking, impedance contrast) will give rise to a strong reflection, and also that if Z_2 is less than Z_1, i.e. if the reflection is at the surface of a slower medium, the reflection coefficient is *negative*. This simply means that such a reflection takes place with a complete reversal of phase: a compression is reflected as a rarefaction, and vice versa.

The ray 2 of Fig. 2.5 gives rise then to two rays: a reflected ray returning to the surface to be recorded at G, and a refracted ray in a direction more nearly parallel to the boundary than was the incident ray.

In Fig. 2.6 another ray 3 of the same pulse front is drawn: this meets the interface at a greater angle of incidence θ_c which is in fact so great that the refracted ray path is parallel to the boundary, that is, $\sin \theta_2 = 1$ and so

$$\sin \theta_c = V_1/V_2 \qquad (2.5)$$

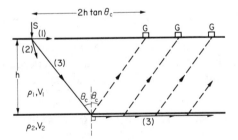

Fig. 2.6   Critically refracted ray path (full line) and a set of rays of the "head wave" pulse (broken lines).

This is the condition of critical incidence and θ_c is known as the *critical angle*: rays meeting the boundary with greater angles of incidence than this are totally reflected. If V_2 is less than V_1 so that the ray path is refracted away from the normal this critical refraction cannot occur.

The ray parallel to the boundary is in the lower medium and the ray treatment alone would lead us to suppose that no energy can be returned to the surface by this path. However, if the theory of wave propagation with these boundary conditions is fully worked out, it can be shown that the pulse, as it travels with velocity V_2 along the lower side of the interface, will generate in the upper medium a pulse of rather small amplitude known as the "head wave" in which the ray paths are inclined at the same angle θ_c to the normal as are the down-going rays which are critically refracted. A geophone on the surface at any distance greater than the *critical range* $2h \tan \theta_c$ from S will lie on one of these rays and will record the arrival of the head wave at the appropriate time.

The nature of the head wave can be qualitatively understood if we remember that the seismic pulse is a region of stress in the rock, and that a pulse in the lower medium will necessarily be accompanied by some stresses in the part of the upper medium immediately above the boundary. This elastically stressed region of the upper medium travels along the interface with the higher velocity V_2 (Fig. 2.7(a)). Since elastic waves can travel in the upper medium only with the lower velocity V_1 the situation is similar to that which exists when an aircraft or rifle bullet travelling at supersonic speed is stressing a region which moves at a higher speed than that of sound waves in air. This results in the generation of a "shock wave" as shown in Fig. 2.7(b), and the "head wave" can be thought of as its analogue. Since the head wave is derived from the refracted wave it is clear that its amplitude at any point should *decrease* with increasing velocity contrast since this increases the proportion of the incident energy returned by reflection.

Although in calculating the travel-time of the pulse by these three paths it is convenient to refer to the ray diagrams of Figs. 2.5 and 2.6, we can better understand the process of its propagation by visualizing the successive positions of the pulse fronts, to which the rays are normal. Pulse-front diagrams for the direct, reflected and refracted fronts, and of the head wave, are given in Fig. 2.8. The reflected front has been shown on a separate diagram merely for clarity: the two diagrams would have to

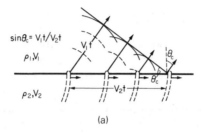

(a)

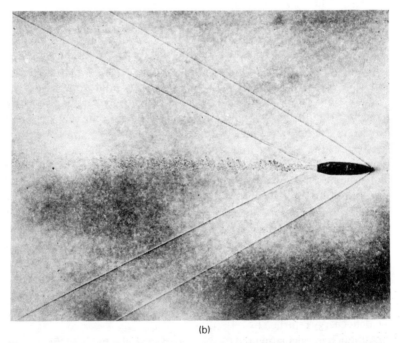

(b)

Fig. 2.7   (a) The generation of a head wave by the passage of a refracted pulse along a boundary. (b) A photograph of the somewhat analogous generation of a shock wave in air by a rifle bullet (from *Sound Waves, Their Shape and Speed* by D. C. Miller, Macmillan, 1937).

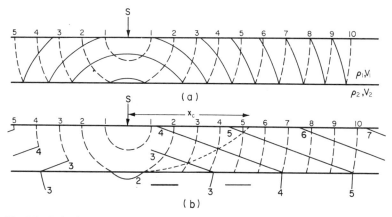

Fig. 2.8 Pulse front positions at equal intervals of time for (a) direct and reflected pulse; (b) direct pulse and refracted and head wave pulses. The dotted line joins points to which the travel-times of direct and head pulses are equal. It reaches the surface at the crossover distance x_c from S. The left-hand part of the diagram has been further simplified by omitting those parts of the pulse fronts which do not produce "first arrivals" at the surface.

be superposed and rotated about the vertical through P to give a true three-dimensional picture of events. It is clear that at a geophone beyond the critical range but not too far from the shot, the direct wave arrives before the head wave, but that at distances greater than a *crossover distance* x_c the head wave, which now spends a greater proportion of its time in the high-velocity medium, is the first arrival. We shall now discuss these ideas in a more quantitative way.

2.3 Travel-time Graphs for a Layered Medium

2.31 *Single horizontal interface*

How will the travel-times along these three paths appear when they are plotted against distance measured along the surface? The graph for the direct P wave has already been given in Fig. 2.3, and Fig. 2.9 is the complete graph for the single horizontal interface of Figs. 2.5 and 2.6, ignoring the slower S and R waves. It is easy to see from the ray diagram

20   *Applied Geophysics for Geologists and Engineers*

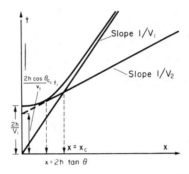

Fig. 2.9   Travel-time graphs for direct, reflected and head waves. A single horizontal interface at depth h divides an upper medium of velocity V_1 from a lower of velocity V_2.

that a geophone near the shot-point will record the arrival of the reflected pulse after a time of $2h/V_1$, and also that at horizontal ranges much greater than the depth h the reflected path becomes practically the same as the direct one, so that the graphs for these two paths must come together asymptotically for large values of x. The actual length of the reflected path is, in fact,

$$2\sqrt{(h^2+(x/2)^2)}$$

as may be seen from Fig. 2.5, and as this is traversed entirely at the velocity V_1, we see that the time-distance relationship for the reflected pulse must be

$$t=\frac{2\sqrt{(h^2+(x/2)^2)}}{V_1}$$

$$=\sqrt{(4h^2+x^2)}/V_1$$

or
$$t^2=(4h^2+x^2)/V_1^2$$

$$=t_o^2+x^2/V_1^2 \qquad (2.6)$$

(since $t_o=2h/V_1$) so that

$$t=t_o \quad \text{for} \quad x=0$$

and
$$t \rightarrow x/V_1 \quad \text{for large } x.$$

The form of the graph for the head wave can also be qualitatively foreseen if we remember that this wave is generated by a disturbance travelling horizontally at the higher velocity V_2 and will therefore have this apparent velocity across the surface: this point is referred to again in the next section. The graph will not, of course, pass through the origin, as does that for the direct wave, since the wave is subject to a delay in travelling down to the high-velocity refractor and to another delay in returning to the surface. These two delays result in an intercept on the time axis if the graph is produced back to $x = 0$. One must realize that the graph has no *physical* reality for angles of incidence less than the critical angle (that is, for values of x less than the *critical range* $2h \tan \theta_c$) as the head wave does not exist at all for such short ranges. Even at slightly greater ranges, the head wave will be so delayed by the inclined parts of its path that it will arrive *after* the direct wave, and only beyond the *crossover range* x_c will the head wave be the first event recorded on the seismogram. Since the head wave amplitude is usually much smaller than that of the direct wave, the part of its travel-time graph before the crossover range can rarely be observed in practice.

The equation of the time–distance graph for the head wave can be found quite readily by adding the times spent on each of the three sections of its path (Fig. 2.10).

Thus

$$t = \frac{1}{V_1} \cdot \frac{h}{\cos \theta_c} + \frac{1}{V_2}(x - 2h \tan \theta_c) + \frac{1}{V_1} \cdot \frac{h}{\cos \theta_c}$$

$$= \frac{x}{V_2} + 2h \left(\frac{1}{V_1 \cos \theta_c} - \frac{\tan \theta_c}{V_2} \right)$$

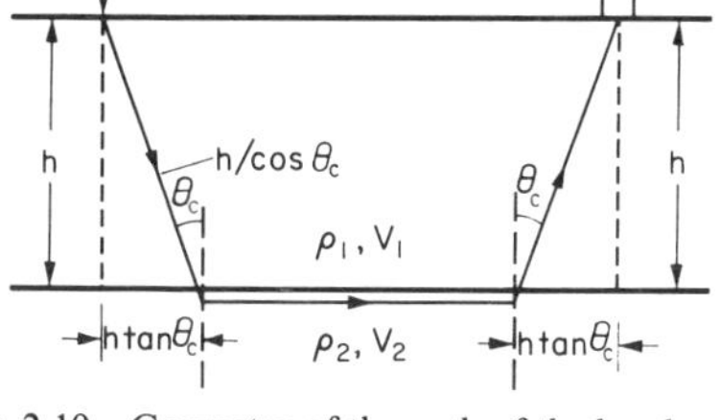

Fig. 2.10   Geometry of the path of the head wave.

Remembering that

$$V_2 = V_1/\sin \theta_c \qquad \text{(from 2.5)}$$

this becomes

$$t = \frac{x}{V_2} + 2h \left(\frac{1}{V_1 \cos \theta_c} - \frac{\sin \theta_c \tan \theta_c}{V_1} \right)$$

$$= \frac{x}{V_2} + \frac{2h}{V_1 \cos \theta_c} \; (1 - \sin^2 \theta_c)$$

$$= \frac{x}{V_2} + \frac{2h \cos \theta_c}{V_1} \qquad (2.7)$$

That is,

$$t = \frac{x}{V_2} + t_i \qquad (2.8)$$

representing a line of slope $1/V_2$ and *"intercept time"* $t_i = 2h \cos \theta_c / V_1$, the intercept of the line on the time axis. The crossover range x_c is the value of x which satisfies both this equation and that of the "direct wave" line $t = x/V_1$, so that

$$x_c/V_1 = x_c/V_2 + t_i. \qquad (2.9)$$

Hence

$$x_c = t_i \left/ \left(\frac{1}{V_1} - \frac{1}{V_2} \right) \right.$$

$$= 2h \left/ \sqrt{\left(\frac{V_2 + V_1}{V_2 - V_1} \right)}. \right. \qquad (2.10)$$

The travel-time equations (2.6) and (2.8) form the bases of the seismic *reflection* and *refraction* methods for determining the depth to an interface. Their use is described in more detail in the next chapter; here we shall only draw attention to the fact that the refraction method relies essentially on the measurement of travel-time of the earliest arrivals at the geophones, and that this is technically a simpler matter than observing reflected waves which occur later in a complex record, for these are of small amplitude compared with the direct wave. This is not to say that later arrivals (in particular the arrivals of the direct wave beyond the critical distance) are of no value in the refraction method: in fact it is always desirable to identify them if possible, and they may often be essential to an unambiguous interpretation.

2.32 *Two or more horizontal interfaces*

Figure 2.11 shows the reflected ray paths and travel-time graphs for a ground with *two* horizontal interfaces. Here two reflected waves are possible (we are neglecting for the moment "multiple" reflections between the interfaces, which are mentioned in Chapter 3) and the minimum travel-time for the second one is

$$t = \frac{2h_1}{V_1} + \frac{2h_2}{V_2} = \frac{2(h_1 + h_2)}{\bar{V}}. \tag{2.11}$$

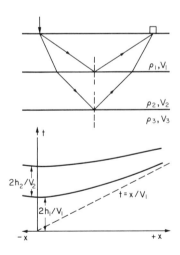

Fig. 2.11   Reflected ray paths and travel-time graphs when two horizontal interfaces are present.

This equation is of the same form as equation (2.6) for $x=0$ if the appropriate average velocity $\overline{V}$ is used where

$$\overline{V} = \text{total path length} \div \text{total travel time}$$

$$= 2(h_1 + h_2) \left/ \frac{2h_1}{V_1} + \frac{2h_2}{V_2} \right.$$

$$= (V_1 t_1 + V_2 t_2)/(t_1 + t_2) \qquad (2.12)$$

so that $\overline{V}$ is on average weighted with the travel times t_1 and t_2 in the layers, the so-called *time-average velocity* which must be used to convert a zero-range reflection time into a corresponding reflector depth. This result is easily generalized to the case of many layers.

For reflections received away from the source where $x \neq 0$, equation (2.6) can still be generalized to the multilayer case by writing it as

$$t^2 = t_o{}^2 + x^2/\overline{V}^2_{RMS} \;(+ \text{ terms in } x^4/V^4, \text{ etc.}) \qquad (2.13)$$

where V_{RMS} is the *root-mean-square velocity* of the layers, defined by

$$\overline{V^2_{RMS}} = (V_1^2 t_1 + V_2^2 t_2 + \cdots)/(t_1 + t_2 + \cdots) \qquad (2.14)$$

and is in fact not greatly different from the time-average velocity.

The graph representing the first arrivals (Fig. 2.12) now consists of three branches whose slopes give the reciprocals of the velocities V_1, V_2 and V_3. Using the method by which equation (2.8) was derived, it can be shown that the intercepts t_1 and t_2 are given by

$$t_1 = 2h_1 \cos \theta_{12}/V_1$$
$$t_2 = 2h_1 \cos \theta_{13}/V_1 + 2h_2 \cos \theta_{23}/V_2 \qquad (2.15)$$

In this notation an angle of incidence θ_{mn} is that of a ray in the mth layer which is critically refracted at the top of the nth layer, so that $\sin \theta_{mn} = V_m/V_n$ by a simple extension of (2.5). This result can also be generalized if more layers are present.

2.33 *Apparent velocity*

Figures 2.11 and 2.12 also serve to illustrate an important concept which is of great use in discussing the interpretation of seismic records,

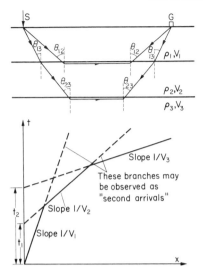

Fig. 2.12   Head wave ray paths and travel-time graphs for two horizontal interfaces.

that of *apparent velocity*, which is simply the velocity of a pulse measured by timing its passage between two closely spaced geophones on the ground surface, that is, the velocity found from the slope of the travel-time graph. For the direct pulse or indeed for any horizontally travelling pulse, the apparent velocity measured along the ground surface is obviously equal to the true velocity V_1, but if the pulse front is inclined at an angle θ to the surface, Fig. 2.13 shows that its intersection with the surface travels a distance $G_1G_2 = x$ while the pulse moves only the smaller distance $x \sin \theta$ along the ray path R. If the true velocity of the pulse is V_1, this will take a time $t = x \sin \theta/V_1$. The apparent velocity measured between G_1 and G_2 will be $V_a = x/t = V_1/\sin \theta$. For the direct pulse, $\theta = 90°$ and, as we have already seen, $V_a = V_1$. For a vertically travelling pulse, such as the reflected pulse (taken to be plane over a small distance) returning to the surface at the shotpoint, $\theta = 0°$ and therefore V_a is infinite. This simply expresses the fact that the pulse arrives simultaneously at any two closely spaced geophones. The apparent velocity of the reflected pulse becomes less and less as it is measured further away from the shotpoint and θ increases to $90°$, when V_a has the limiting value V_1.

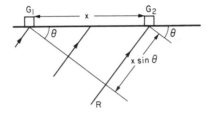

Fig. 2.13   Relationship of true and apparent velocities.

The pulse front of the head wave from a horizontal boundary is inclined at an angle θ_c to the surface, since its ray makes this angle with the normal to the interface, i.e. with the vertical. The apparent velocity of the head wave is therefore constant and equal to $V_1/\sin \theta_c$, and because $\sin \theta_c = V_1/V_2$ it follows that $V_a = V_2$, the velocity of the lower medium, a result that might have been foreseen from the way in which the head wave is generated.

If the interface is dipping downwards in the direction of propagation at an angle a, the head wave will be inclined to the surface at an angle $\theta_c + a$, and the apparent velocity will be

$$V_d = V_1/\sin(\theta_c + a) \tag{2.16}$$

which is *less* than V_2 by an amount depending on a. A dip in the opposite direction will clearly lead to an apparent velocity

$$V_u = V_1/\sin(\theta_c - a) \tag{2.17}$$

which is *greater* than the true velocity of the lower medium. A single travel-time graph for a refractor of unknown dip can give only either V_d or V_u, and to determine θ_c (hence V_2) and a it is necessary to "reverse" the line so that both apparent velocities can be found.

2.4. Seismic Velocity and its Relationship to Lithology and to Other Rock Properties

The seismic interpretation methods to be described in the next chapter result in a "velocity model" of the ground, and if this is to be geologically meaningful it is necessary to relate the velocities to lithology and perhaps also to other rock properties of more practical interest. It is also helpful in planning a seismic survey to be able to use the available geological information to estimate the velocities that are likely to be encountered. For either of these purposes, the formal equations (2.1) and (2.2) are of little value, and indeed could give rise to the misconception that a rock of high density would be expected to have a low seismic velocity. However, rocks are granular materials with average grain densities which do not vary widely so that the bulk density is normally affected more by porosity, that is by the degree of compaction and cementation, than by composition. Elastic moduli of such materials also increase with compaction, depending on the area of contact between grains, and do so more rapidly than does the bulk density, so that the ratios k/ρ and n/ρ which determine the seismic velocities increase with depth of burial in a situation where no changes in composition occur, such as in a uniform sequence of sandstones or shales. Rocks of very low porosity, including most igneous and metamorphic rocks and evaporites, have velocities which are controlled rather by their composition and which can be well predicted from a knowledge of the velocities of their component minerals. Depth of burial is important even in such low-porosity rocks, however, since microcracks may easily reduce the velocity measured at ground surface to little more than half of its expected value. A confining pressure of about 1 kilobar is sufficient to close the cracks, but this corresponds to a depth of burial of 3000 metres.

In rocks of medium to high porosity, the velocity (more particularly the P-wave velocity) will also depend on the nature of the fluid (air, gas, water or petroleum) filling the pore-space so that a schematic velocity-porosity relationship will be as shown in Fig. 2.14. The shallow minimum in velocity at very high porosities corresponds to the stage when grain contacts are minimal and compaction increases the density more rapidly than the elastic moduli. Over the commonly occurring porosity range of 20–40 per cent, the velocity-porosity relationship is quite well approximated by the so-called *time-average equation*

$$\frac{1}{V} = \frac{\phi}{V_f} + \frac{1 - \phi}{V_m} \tag{2.18}$$

in which V_f and V_m are the velocities of the fluid porefiller and of the "matrix" of granular material. This equation is useful for estimating

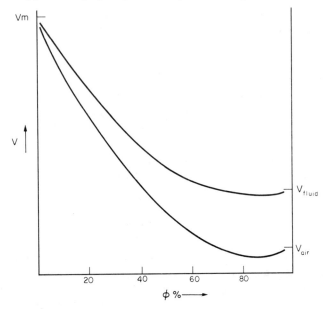

Fig. 2.14   Dependence of *P*-wave velocity on porosity in dry and saturated rocks of grain matrix velocity V_m.

porosity ϕ from velocity or vice versa, using $V_f = 300$ m s^{-1} or 1500 m s^{-1} for dry and saturated rocks respectively, and a value of V_m, appropriate to the composition, which will rarely lie outside the range 5000–6000 m s^{-1}. Partially saturated rocks are found to have velocities close to the "dry" value up to 80–90 per cent saturation, when a rapid increase to the "wet" value occurs.

It is apparent that only ranges of *P*-velocity such as those quoted in Table 2.1 can be expected to correspond to a given lithological type and that the overlap between these ranges is such that a velocity of 4 m ms^{-1}, for example, might correspond to anything from a well cemented sandstone or limestone to a fractured igneous or metamorphic rock. Velocities within one formation, however, normally show a much smaller variability than these figures suggest, and provided that some geological control is available, the identification of such a formation from its seismic velocity in a local context is often quite practicable.

Table 2.1   Approximate *P*-wave velocities for some common geological materials.

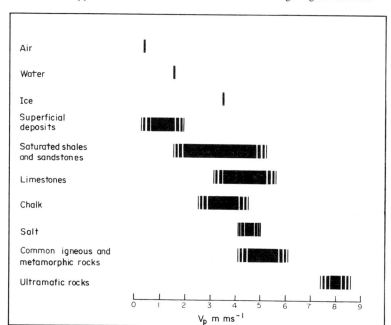

If estimates can be made from velocities of such parameters as porosity, permeability, elastic modulus or strength, they may be of great practical use. We have already seen that moderate porosities can be quite well estimated from velocity provided that it is known whether the rock is dry or saturated. Since rock density depends mainly on porosity, it can also be estimated with a similar confidence, and the empirical curve of Nafe and Drake (Fig. 2.15) is often used for this purpose. The velocity-density relationship for rocks under 10 kbar pressure (equivalent to 30 km depth) has been investigated by Birch (1961) and others. At these pressures porosity is negligible, and the relationship becomes a linear one for a given composition. The most important compositional parameter turns out to be the mean atomic weight, so that if this and the density are known a good velocity estimate can be made without knowledge of the detailed mineralogy of a rock. Velocities measured at depths of less than

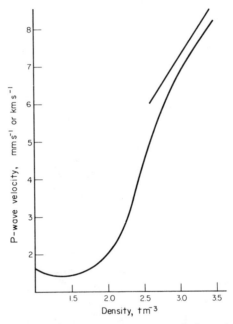

Fig. 2.15   Experimental relationship between *P*-wave velocity and density. (Nafe and Drake, unpublished data, reproduced in Talwani, Worzel and Ewing (1961). Linear section after Birch (1961) for a mean atomic weight of 21.)

3 km are of course likely to be reduced by microcracks below the values estimated in this way.

Permeability is not well estimated from velocity, since although it tends to increase with increasing porosity it is the pattern of connectivity of the pore spaces that is of greatest importance, whilst seismic velocities are reduced by either connected or unconnected pore space.

It can be seen from equations (2.1) and (2.2) that if a rock behaves like an isotropic and perfectly elastic medium, and its density and both body wave velocities are known, the two elastic moduli can be found exactly. These equations can be re-written in terms of any two of the four elastic constants k, n, E (Young's Modulus) and σ (Poisson's ratio) using the relationships which connect them. If only one velocity (usually V_p) is measured, then a value for σ must be assumed before k, n or E can be estimated, for example from the equation

$$\rho V_p^2 = E(1-\sigma)/(1+\sigma)(1-2\sigma) \tag{2.19}$$

Poisson's ratio (the ratio of lateral contraction to longitudinal extension when a rod is stretched) varies between 0 and 0.5 and is commonly taken to be 0.25 for "hard" rocks. For unconsolidated near-surface materials, however, it may reach values of 0.4 or even more, tending towards the limit of 0.5 for a fluid. It can be shown that

$$(V_p/V_s)^2 = (2-2\sigma)/(1-2\sigma) \tag{2.20}$$

so that $V_p = \sqrt{3}\,V_s$ for $\sigma = 0.25$.

In estimating elastic properties from seismic measurements, it must be remembered that the rock is stressed by the passage of the seismic wave within the wave period of a fraction of a second. The long-term stress-strain measurements used in engineering test programmes often give rise to greater strains for a given stress, and therefore to smaller elastic moduli. Moreover, laboratory measurements of "dynamic" modulus carried out by vibrating small specimens free from flaws may give higher values than those based on field velocities which may be reduced by the presence of fractures too widely spaced to be seen in the specimens.

Estimates of rock strength can be made from seismic velocities and can be of practical value if used with discretion. In a given lithology, strength will be lowered from a typical maximum value to an extent which depends on fracturing and weathering and which can be assessed (often indirectly in terms of the number of fractures per unit length) by a field velocity measurement. Ranges of velocity can often be established which are typical of material weak enough to be ripped by mechanical digger or strong enough to require blasting. The bearing capacity of a foundation will be directly measured at many points, but the risk of a weak zone being present between them can be minimized at little cost by interpolating strength estimates from seismic velocities.

CHAPTER 3

Seismic Surveys and their Interpretation

3.1 Introduction

WHEN seismic velocity increases with depth, energy is returned to the surface by refraction and measurements of travel-time as a function of range can in principle be "inverted" or transformed into a graph of velocity as a function of depth. This is the essence of the *seismic refraction method,* whereas the *reflection method* depends on the return of reflected energy from velocity discontinuities which can be either increases or decreases. The simplest *reflection profilers* record these reflections on a single detector close to the source, giving a measurement of the vertical incidence travel time $t_o = 2h/V$ (eqn. 2.6), but the velocity V of the material overlying the reflector has to be known before its depth h can be found. Recording travel-time as a function of horizontal range from the source makes it possible using equation (2.13) to determine V and so to produce a complete interpretation. The reflection method can be extended more easily than the refraction method to deal with the complex multi-layer situations that are common in sedimentary basins, and it is this method that has been particularly extensively developed for petroleum exploration. Its advantages in giving an interpreted cross-section which can be closely related to the real geological section are so plain that it seems at first sight surprising that the refraction method, giving only a simplified ground model consisting of a few major layers, has survived its competition. There are, however, two types of problem which can at present be more successfully solved using the refraction method; firstly, those involving only shallow depths of the order of 100 m and less which are typical of site investigation and hydrogeological work, and secondly the investigation of the deep structure of the earth's crust.

In the first case, the reflected wave returns to the surface so rapidly that it is difficult to separate it from other events in the early part of the seismogram, a difficulty which could be overcome by using a source of higher frequency than the normal. This is only practicable in water-covered areas (see §3.2). In crustal investigations the difficulty is a more fundamental one: velocity contrasts in the deep crust are small, and the laterally continuous layering characteristic of sedimentary basins is rarely found, so that velocity boundaries are often gradational vertically and impersistent laterally, giving rise to short unrelated reflecting segments rather than to continuous horizons. We consider in this chapter only the procedures for shallow (< 100 m) refraction work and for conventional reflection prospecting which covers the depth range from a few hundred to a few thousand metres.

3.2 Seismic Sources

The source should ideally provide a pulse of duration of no more than a few milliseconds and large amplitude and should be safe, cheap and repeatable. All these requirements are reasonably well fulfilled by the small explosive charge fired in a hole up to a few tens of metres deep which was used almost exclusively in the early days of seismic prospecting, but a wide variety of non-explosive sources has now been added to the conventional "shot". They are conveniently divided into those used on land and in water-covered areas.

On land, explosive charges are still quite commonly used in reflection surveys and for refraction work involving surface ranges greater than 50–100 m, corresponding to depths of investigation of over 10 m or so. They give a good source pulse of high frequency and amplitude but require several light drills to work with each recording party if steady production of data in a reflection survey is to be maintained. Drilling of shot-holes may become impracticable if access is difficult or the surface layers present drilling problems, and in this case one of the various *surface sources* may be chosen in preference to explosives. These all generate seismic waves of small amplitude (which is an advantage when working in populated areas) and therefore could not be used widely until the advent of magnetic recording (see §5.3) made it possible to sum or *"stack"* a number of seismograms from repetitions of the source at one point to produce a net effect comparable with that of a single explosive shot. *Weight-drop* sources in which a weight of about 10 kg strikes a surface coupling plate after falling some 3–4 m are often used in site investigation surveys for depths up to about 10 m: a sledgehammer with an enthusiastic operator can give comparable seismic energy. For deep reflection prospecting, weights a few hundred times greater are dropped through about the same height, and many drops are made at the same or neighbouring points for stacking in the recording process (Neitzel, 1958). Sources of the *gas-gun* ("Dinoseis"*) type are also used in reflection work: here the impulse applied to the coupling plate comes from the explosion of a propane–oxygen mixture in a heavy chamber attached to it. *Explosive cord* ploughed in just below the surface has been found effective when conventional shooting is difficult, and has some advantages in safety and ease of handling. These surface sources, particularly the weight drop and gas-gun, permit much more rapid repetition of "shots" than does conventional shooting, but still suffer from the fundamental limitation that they produce a seismic pulse with a form controlled by source energy (large energy leads to a long duration "low frequency" pulse) and near-source ground properties. To obtain a pulse of any desired waveform requires a vibrator as the source, and the "Vibroseis"† system described by Crawford, Doty and Lee (1960) uses a vibrator controlled electromagnetically or hydraulically to produce a "swept-frequency" waveform extending over 10 seconds or more which usually starts as a sinusoid of frequency about 6Hz, this frequency being increased continuously to perhaps 60 Hz so that the whole frequency band recorded in reflection seismology is covered. The reflected signals will of course overlap each other, since reflection travel-times are commonly no more than three or four seconds, and the seismogram produced is therefore unintelligible until these signals are "compressed" in time by a process of cross-correlation with the known source signal. This process effectively searches for the repetitions of the source signal contained in the seismogram, giving a pulse of short duration (which is in fact the "autocorrelation function" of the source waveform) each time such a repetition, or reflection event, is found. The seismogram after

*Trademark of Atlantic Richfield Co.
†Trademark of Continental Oil Co.

cross-correlation is thus comparable with one obtained from an explosive source. This system is now extensively used on land, even in built-up areas since the disturbance produced by the source is negligible.

Marine sources are even more varied and no more than a selection can be described here. An explosive charge is much more efficient in producing seismic energy when fired in water than on land, but considerations of safety and convenience have led to the almost complete replacement of explosives by *air-guns* in marine reflection surveys for petroleum. Such guns release a bubble of high-pressure air when the movement of a piston is triggered electrically, and are towed in an array behind the recording ship. The total seismic energy released by the array is comparable with that of an explosion, and the waveform can to some extent be controlled by combining guns of different sizes which can produce dominant frequencies in the range 10–100 Hz.

The bubble which is produced by both air-guns and explosive sources oscillates as it rises to the surface at a frequency dependent on source energy and depth. The seismic wave generated therefore consists of a primary source pulse and a train of "bubble pulses" which confuse the seismogram. Most of the variants on marine impulsive sources are designed with the aim of eliminating the bubble pulses either by dispersing the bubble, preventing its collapse, or causing an implosion rather than an explosion for the primary pulse. Some of these variants still use small explosive charges as the energy source.

Marine refraction surveys are now limited entirely to the "crustal" scale involving ranges of tens to hundreds of kilometres for which only explosives are suitable.

Small scale, high resolution seismic surveys in shallow water are important in site investigation and in the search for resources such as gravel deposits. They can usually be carried out by the reflection method, since the negligible attenuation of seismic waves in water means that higher frequency sources can be used than is possible on land. The usual attenuation increasing with frequency (§2.2) limits the sub-bottom penetration, and a compromise must be adopted between increasing the source frequency (shortening the wavelength) to improve layer resolution and decreasing it to achieve deeper penetration. Some common sources for work on this scale are the "sparker" (condenser discharged across underwater spark gap, frequency $f \simeq 100$–400 Hz), the "boomer" (metal disc spring-mounted against a coil from which it is sharply repelled by the passage of an impulsive current, $f \simeq 200$ Hz) and the magnetostrictive transducer also used in echo-sounders and sonar ($f \simeq 5$ kHz for useful penetration). The resolution for each of these will be of the order of half a wavelength in near-bottom sediments: say a few metres for sparker and boomer or a fraction of a metre for the transducer. Sonar-type transducers are also used in the *side-scan sonar* system in which a narrow beam of pulsed high-frequency sound is directed sideways from a moving ship giving returns from scarp-like features of a rocky bottom which can be used in geological mapping. If the bottom is sediment-covered the returned energy is scattered and is of greater amplitude from coarser sediments, so that gravels can readily be distinguished from sands or muds.

3.3 Recording Systems

A complete seismic recording system or *seismograph* consists of elements for detecting, recording and displaying the motion resulting from the arrival of a seismic wave: these elements will be considered in turn.

3.31 *Detectors*

On land, the ground motion is detected by a *seismometer* or *geophone*, which is an electromechanical transducer commonly of the moving-coil type shown in Fig. 3.1. The coil is free to move in the annular gap between the pole-pieces of a permanent magnet, and the output voltage is proportional to the velocity of this relative motion, which is identical to the motion of the case (i.e. the ground motion) for frequencies well above the resonance of the coil suspension. Below this resonance frequency the coil and case tend to move together and the output falls off rapidly as indicated in Fig. 3.2, which also shows how the damping of this resonance affects the response. It is clear that the resonant frequency of a seismometer must be chosen to be below the lowest frequency that one wishes to record. The "planting" of a seismometer on the surface can also alter its effective response and its sensitivity to wind-generated "noise"; it may be helpful to remove loose surface material and drive the

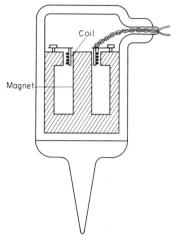

Fig. 3.1   Simplified sectional view of a typical moving-coil geophone.

spike into firmer ground, but routine reflection work relies on large numbers of geophones rather than on careful individual placement.

Water-borne surveys normally use a pressure-sensitive *hydrophone* as the detector, which may be suspended floating freely below the surface in refraction surveys, but in reflection work is incorporated in large numbers into a tubular *streamer* or "snake" one or two kilometres in length, which is towed behind the recording vessel about 10 m below the surface at a speed of about 10 km/hr.

In principle each geophone (or hydrophone) detects the ground motion at one surface point and this information is passed through a single recording *channel* and appears as one trace on the resulting seismogram. In practice, in reflection surveys, many detectors may be connected together as a *group* so that it is their summed outputs that pass through the recording channel. The group may cover up to 100 m and the detectors in it are so arranged that its output emphasizes the reflections arriving on near-vertical ray paths at the expense of unwanted events such as surface waves. In what follows we shall think of the group as equivalent to a single detector at its centre, and shall refer to "geophones" whilst not excluding hydrophones.

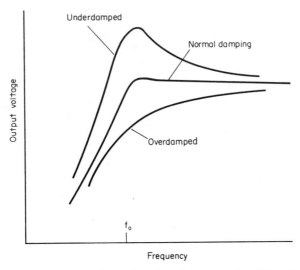

Fig. 3.2 Output of a geophone as a function of frequency for different degrees of damping. f_o is the resonance frequency.

Simple single-channel systems are sometimes used in small-scale refraction work when the source or recorder is very mobile. If the single geophone is left at one point, the time-distance graph can be built up by moving the source (e.g. a weight-drop or hammer-blow) to successively longer ranges. The normal-incidence reflection profiler has already been mentioned (§3.1) as another example of a useful single-channel system, but in general for either refraction or reflection work it is most efficient to gather seismic data in *multi-channel* form, and 12, 24, 48 or even 96 channels are now in common use.

The detectors of such a multi-channel system are usually arranged along a straight profile at right angles to the general strike for simplicity of both operation and interpretation, but the seismogram may include confusing events from reflectors with a component of dip across the profile. In surveys where three-dimensional information is essential, a grid of intersecting profiles will be laid out.

3.32 *Geometry of refraction profiles*

Spacing of detectors along a seismic refraction line is determined by the detail in which one wishes to examine the refracting horizon, since a depth estimate can in principle be made at each geophone position. Once this spacing has been decided, the ''spread'' length which can be covered by one shot is decided by the number of channels in the system, which is usually the same as the number of pairs in the multicore cable used to connect the geophones. The length of a complete refraction line, however, must clearly be at least several times greater than the cross-over range x_c (eqn. 2.10) and in planning a survey this minimum line length is conveniently taken to be about 10 times the estimated depth to the horizon of interest. If the required line length exceeds the spread length, the time-distance information has to be built up from several shots as indicated in Fig. 3.3. Continuous coverage of the refracting horizon can then be obtained by overlapping any number of such lines to form a profile.

3.33 *Geometry of reflection profiles*

The principles governing layout of detectors in reflection work are rather different: the maximum spacing between geophones (or rather between group centres) is still determined by the coarseness of the horizontal resolution that can be accepted, but a more stringent requirement for close channel spacing is likely to be set by the need to recognize and correlate the reflection events in the confused later part of the seismogram, a need which arises in refraction profiling only if ''second-arrival'' events are used in the interpretation. It is clear from inspection of the isolated trace above Fig. 3.4 that reflections will only be seen if they correlate well across a number of channels of the seismogram, and the variability of reflecting horizons is such that the spacing between geophone groups cannot in practice exceed about 100 m. As before, the number of channels n determines the overall spread length as $(n-1)d$ for a group interval of d, but the geometry of the reflected ray path implies that the *subsurface cover* is only $(n-1)d/2$ with the points to which depth estimates are made or *depth points* being separated by half the group interval: see Fig. 3.5.

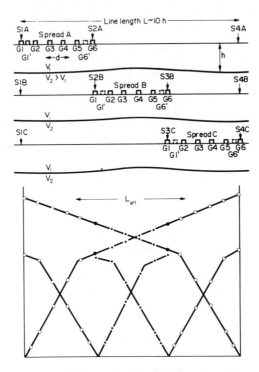

Fig. 3.3   Arrangement of spreads and shots in a refraction line using 6-channel equipment. Estimates of the depth to the refractor are required at a spacing of d. The geophones G_1, G_6 are moved to G_1', G_6' when a shot is fired at their normal position. The schematic time-distance graph shows the information obtained from each of the spreads A, B and C. The four short-range shots at shotpoints 2 and 3 are to determine whether the overburden velocity V_1 changes along the line. "Reversed" coverage of the refractor (see §2.33) is obtained only over the length L_{eff} between the crossover ranges from S_1 and S_4, and the end-point of the next line to the right would therefore have to be S_3 rather than S_4 to obtain continuous reversed coverage.

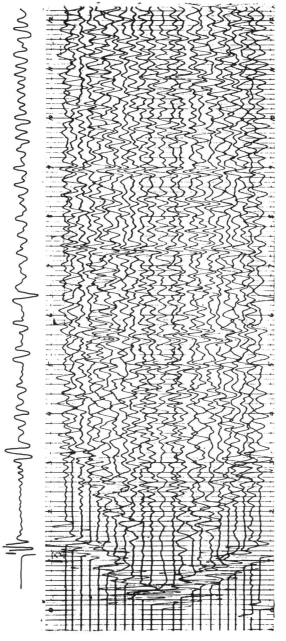

Fig. 3.4 A typical reflection seismogram. The 24 geophones are spread at 30 m intervals up to 330 m on each side of a central shotpoint. If the end geophones coincide with the shotpoints of adjoining geophone spreads, continuous coverage of the reflecting horizons is obtained (see Fig. 3.5).

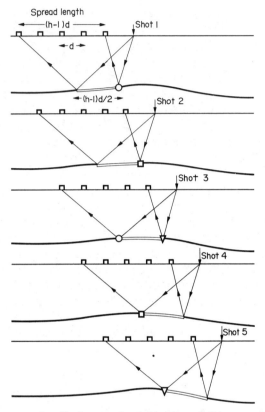

Fig. 3.5   Movement of a reflection spread and shotpoint to build up a profile with double subsurface coverage. Shots 1, 3 and 5 alone would give simple continuous coverage as the spread is moved by half its length between them. The constant range from shotpoint to nearest geophone is arbitrary but is usually no more than a few times the group spacing d. The points marked O, □ and ∇ are common depth points on the reflector for shots 1 and 3, 2 and 4, 3 and 5 respectively.

If the length of a reflection line is either too long (leading to breakdown of approximations of the simple theory) or too short (giving a small difference between t and t_o in equation 2.13), the determination of the overburden velocity, and hence of the depth to the reflector, from equation (2.13) becomes inaccurate. The compromise generally adopted is to use a line length (often the same as the spread length) of the same order as the reflector depth of greatest interest, and one of the many advantages of the reflection method is that this line length is thus some ten times less than that required for a refraction investigation of the same horizon. To join reflection lines together in a profile giving continuous subsurface cover clearly requires an overlap of half the spread length which is provided by shots 1, 3 and 5 of Fig. 3.5. For a movement of spread and shot of only one-quarter of the spread length, shots 2 and 4 duplicate coverage of reflector so that two record traces are obtained from each depth point. This principle is developed in the *common depth point method* of shooting a reflection profile in which spread and shotpoint are moved between shots by an even smaller fraction of the spread length to give multiple coverage of the reflector which may be 6, 12 or 24-fold. The procedure is particularly easy in marine work with an air-gun or similar source, simply requiring more frequent firing of the gun as the hydrophone streamer is towed along the profile. On land the shotpoint is normally moved by some multiple of the group spacing and geophones to cover several spreads which are planted simultaneously, with a different set of geophones being switched into the recording channels for each shot. The method is discussed further in §3.4; multiple coverage makes it possible to reinforce the weak reflection arrivals and has been one of the most important advances in reflection technique.

3.34 *Recorders and display systems: refraction surveys*

The emphasis in shallow refraction surveys is on obtaining accurate measurements of first-arrival travel times, so that a simple multi-channel amplifier and galvanometer recorder giving a paper record of the ''wiggle-trace'' type is all that is normally required. The amplifiers have individual gain controls which can be set to give the highest gain allowed by the background seismic noise, and usually some form of filtering to exclude frequencies outside the useful signal range of perhaps 20–200 Hz. The recording is on ultraviolet sensitive paper which requires no processing other than exposure to diffuse daylight. The ABEM ''Trio'' shown in Fig. 3.6(a) is an example of such a system, and Fig. 3.6(b) is a typical refraction record on which the timing lines, in this case at 2 ms intervals, and the pulse derived from the firing of the source (the ''shot-instant'') can also be seen.

Fig. 3.6(a)   The ABEM "Trio" 12-channel refraction recording system.

Other display systems are also available, particularly for simple single-channel seismographs, in which the result of a single shot may, for example, be seen on an oscilloscope screen and the first break timed using a movable calibrated marker pulse, or coded as a series of marks made by a stylus on electrosensitive paper. Another variation dispenses with the display altogether and reduces the system to an *interval tuner*, an electronic clock started at the shot instant and stopped by a pulse derived from the first arrival at the detector. These simpler systems may have

Fig. 3.6(b) (*opposite)*   A typical refraction record.

45

advantages in cost and portability over those with many channels and a permanent display, but there is greater difficulty in correlating arrivals and ensuring that a weak first arrival is not missed. This latter difficulty is minimized in recorders (e.g. the "Bison" and E.G. and Co. "Nimbus") which incorporate a digital summing device enabling the seismograms from repeated hammer or weight sources at the same shotpoint to be added before display on an oscilloscope and digital measurement of the travel-time.

Refraction investigations on the "crustal" scale use individual self-contained seismographs, usually recording on magnetic tape, spaced at intervals of the order of a kilometre and receiving timing and shot-instant pulses by radio.

3.35 *Recorders and display systems: reflection surveys*

The reflection seismograph has of necessity evolved into a much more sophisticated instrument than those just described: the amplifiers must be able to deal with the wide range of amplitudes or *dynamic range* between the early arrivals by the direct path and the weak reflections from depth; and these reflections are often difficult to distinguish from the various unwanted events which are classified as "noise". Improvement of signal-to-noise ratio can be obtained by careful choice of frequency-selective filters and by adding together (after "moveout" correction: see §3.42) the many seismograms obtained from one point on the reflector by the common depth point procedure. Simple recording on paper permits neither second thoughts about optimum filter characteristics nor summation of records, and so reflection records are now always made on magnetic tape, which can be replayed into a data processing system where a wide range of filtering operations including time-shifting (see §3.42) and summation or "stacking" can be performed. The magnetic recording used from about 1955 to 1965 was of the "analogue" or continuous type, but since that time there has been a rapid transition to *digital recording* which is now universally favoured. Each trace of the seismogram is instantaneously "sampled" at intervals which can be from 1/4 ms to 4 ms depending on the resolution required. The amplitude values are written to computer tape: such sampling can give faithful recording of amplitude ratios of up to $10^5:1$ and frequencies up to about 1000 Hz.

When digital recording is used, a paper record is made of selected shots to monitor data quality, but the display used for interpretation is produced by replay of the traces after all necessary data processing has been carried out. The traces from a whole profile are replayed side by side on a reduced scale to form a *seismic section*: for an example see Fig. 3.7. Each trace uses the "variable area" presentation in which negative loops are omitted and positive loops filled in, making reflection events particularly easy to correlate.

In single-channel reflection profiling systems, digital recording is less advantageous, but the hydrophone signal is sometimes magnetically recorded in analogue format for subsequent refiltering. The display is directly in the form of a seismic section, using some form of facsimile recorder which draws a trace about 1/4 mm wide for each shot, with light and dark shades corresponding to the negative and positive excursions of a conventional record. Figure 3.8 is an example of a profiler section in this "variable density" presentation.

3.4 Reflection Surveys

3.41 *Field procedure*

On land the nature of the terrain, whether a built-up area or farming country; desert, mountains, swampy or permanently frozen ground, will obviously determine the tactics used by a seismic contractor to complete a seismic profile, and the way in which the preliminaries of permissioning and surveying are carried out. The position and spacing of the seismic profiles will however be dictated by the broader strategy of the client, usually an oil company, which is based on existing knowledge of the geology of the sedimentary basin. Initially this knowledge may be only of the extent and approximate thickness of sedimentary rocks, estimates of which may have been aided by gravity and aeromagnetic surveys, but even a few preliminary seismic lines may give some idea of the structures and types of petroleum "trap" that may be present. The simple anticlinal trap may not need many more profiles to define it sufficiently well, but traps involving faults or "pinchouts" of sedimentary layers will usually require much more seismic work to map them properly.

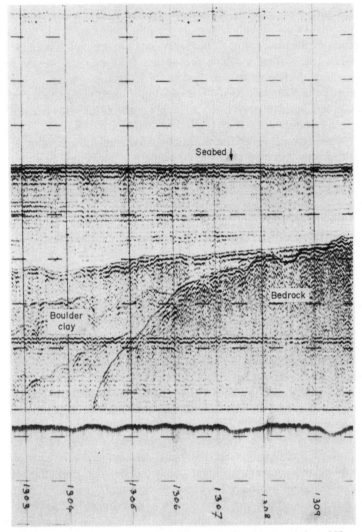

Fig. 3.8   A "boomer" record from the Irish Sea, reproduced by permission of Electronic and Geophysical Services Ltd. Stratified post-glacial sediments overlie boulder clay which is banked against bedrock to the right. A multiple of the sea-bottom reflection is also seen, cross-cutting the sub-bottom reflectors. The time zero is above the top of the picture, and the timing lines are about 8 ms apart. The horizontal scale is marked in 100 m intervals.

The spacing between centres of geophone groups is normally in the range 50–100 m, but the geometry of the groups themselves is decided after the wavelength of the unwanted surface waves has been found from trial records. If dynamite is being used, an optimum charge size (2–20 kg) and hole depth (2–50 m) will also be found by trial and error. For any type of source, a pattern of "shot" points may be chosen to discriminate by summation against the recording of surface waves. Frequency filtering is used only to eliminate very troublesome noise frequencies, the normal practice being to record over as wide a band of frequencies as possible and to leave all filtering to the data-processing stage.

Common depth point coverage is achieved by planting as many geophone groups as cabling permits, connecting the recording truck to the cable at some point and using a "roll-along" switch to connect to the recording channels the set of groups appropriate to the shot being fired. The "live" groups thus move one shotpoint interval along the cable as the shotpoint is moved, and the process continues until the "live" groups are at the far end of the cable, when its rear section must be picked up and added to the advancing end.

A land reflection crew may include about twenty people (though as many as 500 can be involved in terrain such as the Niger delta) and should be able to cover some 5–10 km of 24-fold coverage per day, though this figure can of course vary widely.

Recording at sea is similar in many ways to work on land, though the problems of access and permissioning are replaced by those of accurate navigation, now controlled by such aids as precise radio beacons, Doppler sonar, and satellite fixes. The procedure is even more of a set routine than in land recording, since the length of the streamer makes any change in the ship's course slow and therefore too costly to be justified except when the monitor records show that very poor data are being obtained. The streamer depth (10–20 m) is carefully controlled to avoid changes in the inevitable interference between signals reaching the hydrophones directly and those reflected downward from the sea surface. The position of the end of the streamer is also monitored as currents will often deflect it from the line of the profile. Common depth-point recording is simply carried out by firing the source, which is close to the ship, every time the streamer has advanced by n hydrophone spacings. If there are N groups on the streamer, this will result in F-fold coverage where F is the next integer above $N/2n$.

Marine recording can cover profiles about ten times more rapidly than is possible on land, and since daily costs of a marine crew are not greatly different, its production costs per kilometre may be an order of magnitude smaller than for land surveys.

3.42 *Data processing*

Single-channel profiler systems give a display which is directly in the form of a seismic section of closely spaced individual seismograms (Fig. 3.8), but multi-channel data of the common depth-point type have to undergo a complex sequence of corrections and other processing before the section is produced.

The first correction is known as the *datum, weathering* or *static correction* and is of particular importance in land surveys. It allows for the effect on reflection travel times of irregularities in surface topography and of the thickness and velocity of the *low-velocity layer* (LVL) or *weathered layer* which extends to depths of up to tens of metres. The object of the correction is to reduce the travel times to those that would be observed if shots and geophones were placed on a *datum surface*, usually horizontal, passing below the lowest point of the LVL: this is a simple matter if the form of the ground surface and the thickness and velocity of the LVL are known everywhere along the profile. The required information about the weathered layer may be obtained directly by *uphole shooting*, that is by drilling selected shotholes to well below its base and firing a series of charges at various depths within them, recording travel times to a shot geophone spread near the top of each hole. Between these direct measurements, delay times in the weathered layer may be estimated from the head waves generated at its base and recorded as first arrivals on the reflection seismogram: travel times of these first arrivals can be interpreted by the methods of §3.5. A different approach to the static correction is to suppose that it is the sole cause of irregularities in the travel times of a shallow reflecting horizon, and to simply apply those time shifts to the traces which most effectively smooth the reflector. This method can be elaborated statistically, and is satisfactory if the chosen reflector is indeed smooth, an assumption which can be checked by the effect of the correction on the smoothness of deeper reflectors.

The seismograms are next rearranged by computer from the "common shotpoint families" in which they are arranged on the digital tape produced by the recorder into "common depth point families". For example, in Fig. 3.5, the seismograms recorded by shot 1 form a common shotpoint family, whereas a common depth point family is formed by selecting the record of shot 1 at the first geophone, shot 3 at the fifth, and so on to more distant geophones not included in the figure. The ray paths corresponding to such a CDP family will be those shown in Fig. 3.9 and

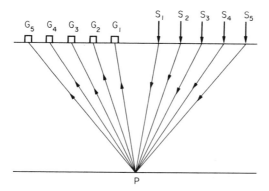

Fig. 3.9 The ray paths which form a common depth point family from a horizontal reflector. The five geophones are shown in the positions that they occupy when shot 1 is fired: for S3 the spread has moved two spacings to the right so that it is G5 which now occupies the "G3" position.

for a surface range or offset distance of x the travel time t of the reflected pulse is given by equation (2.6) as

$$t^2 = t_o^2 + x^2/V^2$$

where $$t_o = 2h/V$$

and the material above the reflector has a constant velocity V which can in principle be determined by plotting t^2 against x^2. The increase in travel time due to the offset, $t - t_o$, is known as the *normal moveout* or NMO and can be found by writing

$$t^2 - t_o^2 = x^2/V^2$$

as $$(t - t_o)(t + t_o) = x^2/V^2$$

and supposing that $t - t_o$ is small (as it is for $x \ll h$) so that

$$t + t_o \simeq 2t$$

Hence $$\text{NMO} = t - t_o = \Delta t \simeq x^2/2tV^2 \qquad (3.1)$$

Note that the NMO decreases with increasing t, that is, the hyperbolic $t - x$ relationship for a reflection event (Fig. 2.9) is less strongly curved for the deeper reflectors. A correction for NMO will thus depend on travel-time on a single seismogram and is sometimes called the *dynamic* correction in contrast to the static correction which has the same value for the whole of any one trace.

If each seismogram of a CDP family is corrected for moveout using the value of Δt calculated from x, t and the correct velocity V, any reflection event will align at its zero-offset travel time t_o instead of lying on the usual hyperbola (Fig. 3.10) whereas of course arrivals other than reflections (i.e. noise) will not be brought into alignment. If the alignment is tested by the obvious method of summing or *stacking* the traces, the amplitude of the reflection event will be increased n times if n traces are stacked, but since random noise would be increased in amplitude only $\sqrt{n}$ times, the *signal-to-noise* ratio (S/N ratio) will be increased $n/\sqrt{n} = \sqrt{n}$ times. This improvement can only be achieved if the

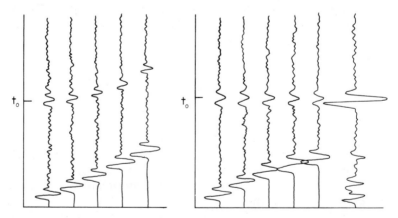

Fig. 3.10   Simplified set of CDP seismograms showing the effect of the normal moveout correction on a reflection and on the "first breaks" (direct arrivals). The sum of the five corrected traces is also shown on the right.

reflections are perfectly aligned: if the NMO correction is calculated using various velocities differing slightly from the true value, and the stacking is repeated, it is clear that the greatest reinforcement of the amplitude of the stacked reflection will occur for the true value V, giving a method of estimating the velocity of the material overlying a reflecting horizon. This is the method commonly used for estimating velocities in reflection profiling, having the advantage of being only a variation in the stacking process which is an integral part of the CDP technique. Older methods such as plotting t^2 against x^2 or t against Δt are useful for individual seismograms but cannot be incorporated into a continuous production routine.

As we have seen in §2.32, equation (2.6) can be extended to the case when many layers overly the reflector of interest and becomes

$$t^2 = t_o^2 + x^2/\overline{V^2} \qquad (2.13)$$

so that the velocity for optimum stack of the reflection or *stacking velocity* is approximately the RMS velocity of the overburden. If the RMS velocities for a sequence of reflecting horizons are found in this way, the *interval velocities* of the individual layers V_1, V_2, etc. can be found by successive applications of equation (2.14) from the shallowest reflector downwards. Thus

$$\overline{V_1^2} = V_1^2$$

so V_1 is known.

$$(t_1 + t_2)\overline{V_2^2} = V_1^2 t_1 + V_2^2 t_2$$

in which all quantities are now known except V_2: and so on.

The summed trace is drawn as one element of a seismic section in which the vertical scale is still travel time and the horizontal represents position along the profile, each trace being placed at the central point of the set of shots and detectors which have contributed to it. Reflection events in this *vertically plotted* section thus appear vertically below the corresponding point of observation at their zero-offset travel times. This is as it should be when the reflector is horizontal, but Fig. 3.11 shows that for dipping horizons the situation is not quite so simple. The common depth point P is not in fact vertically below O, the mid-point of the shots and geophones of the family, but is displaced in the up-dip direction by a distance which increases with dip. The vertically plotted section will thus show anticlines as broader and synclines as narrower than they really are, and this distortion of the section, which can make it difficult to interpret when strong folding or faulting is present, is commonly removed by the process called *migration*. This was originally carried out manually on picked reflector segments but is now a computer process applied to the whole section. It requires a knowledge of the velocity distribution as does the conversion of the vertical axis from time to depth which may follow it or be combined with it.

Migration also removes the convex hyperbolic events known as *diffractions* which originate from point or line scattering centres such as faulted edges of reflecting horizons, since all segments of such an event migrate back to the source point.

The reflection section will often contain unwanted information in the form of *multiple reflections* coming from ray paths such as those shown in Fig. 3.12. Some of these will obviously be easily recognizable from their simple time relationships to the primary events, and a primary at reflection time $2t$ may be distinguishable from the double reflection of the horizon at time t by its moveout, which will be appropriate to the greater value of $\bar{V}$ normally found at greater depth. Multiple events corresponding to seismic energy rebounding between a closely spaced pair of good reflectors is known as *reverberation,* the most common example being that set up between the surface and the sea-floor in marine prospecting. In shallow water such closely spaced multiples can give the seismic section the appearance of having been "tuned" by passage through a filter with a narrow pass-band of frequencies and their removal

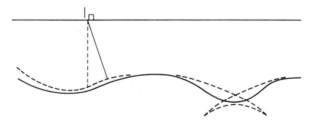

Fig. 3.11(a)   Schematic relationship between true and vertically plotted (i.e. "migrated" and "unmigrated") positions of a reflector including a gentle syncline and anticline, and another syncline with a radius of curvature smaller than its depth below the surface.

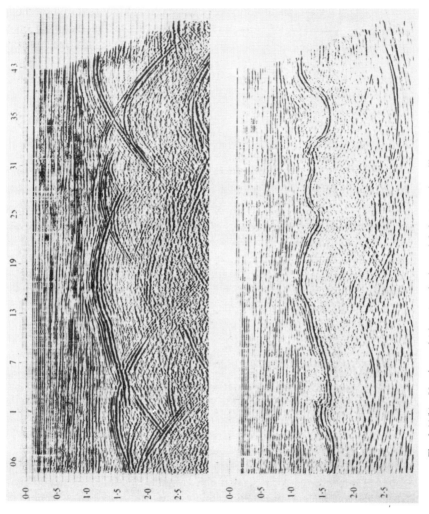

Fig. 3.11(b)   Unmigrated (above) and migrated (below) sections illustrating the effects of Fig. 3.11(a).   Reproduced by permission of AGIP and Western Geophysical from Telford *et al.* (1976), Fig. 4.106.

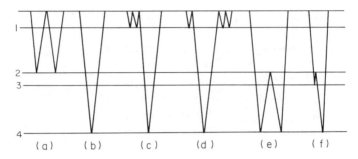

Fig. 3.12   Examples of some ray paths of multiple reflections. Paths (a) and (b) might have identical travel times but could be distinguished on moveout if the r.m.s. velocities to reflectors 2 and 4 were sufficiently different. (c) and (d) are examples of reverberation in a near-surface layer: for example reflector 1 might be the seafloor in a marine survey. (e) and (f) are examples of "long-path" and "short-path" multiples.

("dereverberation") is a process of *inverse filtering* or *deconvolution* which has only become possible since digital filtering has replaced the old analogue methods. Deconvolution in a broader sense has also been applied to removing the effect of the low-pass filter which represents the frequency-dependent attenuation of the earth (§2.2), thereby "sharpening" the broadened impulses which form the seismogram and improving its resolution.

3.43   *Interpretation of reflection sections*

The modern digitally processed section is often of such clarity and obvious geological significance (Fig. 3.13) that it is tempting to interpret it as though it were literally a geological section. This it is certainly not, and the interpretation can only be successful if it draws on an understanding of the principles both of geology and of reflection seismology. Even if the processes of migration and dereverberation perfectly achieved their aims, which is rarely the case, the section may contain "spurious" events other than diffractions and short-period multiples, as, for example, reflections from horizons off the line of the section but dipping towards it. Such events, if they can be identified, are ignored when prominent reflectors are traced on the section or on an overlay as the first stage of

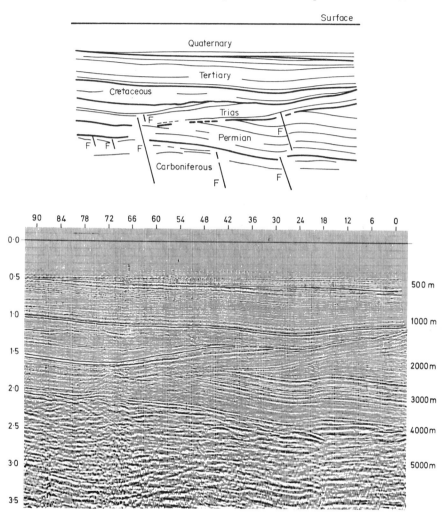

Fig. 3.13   A processed seismic section and its geological interpretation. Reproduced with permission from Anstey (1970), pp. 91–92.

interpretation. The next stage is to assign velocities or velocity gradients to the layered model of the ground indicated by the tracing, using interval velocities derived from RMS (stacking) velocities (eqn. 2.14 and §3.42) or observed in continuous velocity logs (§9.4) from any wells on or near the line. Unless the control on velocity is very good, conversion of the time section to a depth scale is hazardous, and many of the "pitfalls" in interpretation (Tucker and Yorston, 1973) are connected with the imperfect knowledge of velocities that is usually available. This leads not only to incorrect predictions of absolute depth to reflecting layers but, more seriously, to uncertainties in relative depths, that is in the structure which the method is supposed to determine. For example, a high-velocity bed which varies in thickness laterally will give reduced travel times to deeper reflectors below its thickest part, and if the high-velocity bed is not itself identifiable, these times may be interpreted as a significant local reduction in the depth to the lower horizons.

Faults are often difficult to identify on a reflection section: the ideal of an identifiable reflection showing a discontinuous displacement is only rarely observed. More commonly, a group of reflectors show poor continuity in the neighbourhood of a fault, and may be so similar in character that an individual event cannot be correlated across this zone to determine the throw of the fault. The presence of a fault may also be indicated by a series of diffraction events along the fault plane (§3.42) or by the failure of a reflector followed round a loop of profiles to close on itself, as a result of a miscorrelation across one or more faults.

The problems of interpretation are treated in detail by Fitch (1976), McQuillin *et al.* (1979) and Payton (1977).

3.5 Refraction Surveys

3.51 *Field procedure and data handling*

The ideal geometry for a refraction survey has already been discussed in §3.32, and the actual procedure depends on the depth of investigation and lateral resolution that are expected. For a depth of the order of 10 m, we require a line length of about 100 m and a weight-drop source may be adequate in quiet conditions. The delay times to be measured will be perhaps 10–20 ms, so that for a 10 per cent precision in depth estimation we need to measure arrival times to better than a millisecond. A depth of 100 m calls for a more powerful source, usually up to a kilogram of explosive in a shothole 2–3 m deep, to reach a range of about 1 km. Delay times of 50–100 ms can be determined with rather better precision than on the smaller scale, though the predominant period of the seismogram may be 20 ms rather than 10 ms, so that it is still necessary to pick arrival times with an accuracy of a fraction of a cycle. To do so presents no problem if the source energy is great enough to give a very sharp "first break" on the seismogram, but this means that the later parts of the record, where amplitudes are greater, are normally quite unreadable, so that no use can be made of the later arrivals which may carry important information, as we discuss in §3.52. The use of digital recording in refraction work has overcome this limitation of older systems.

The travel times picked are corrected to a near-surface datum and plotted against range, when the points are usually found to fall on a set of well-defined straight-line branches which can be interpreted as follows.

3.52 *Refraction interpretation*

Let us take as an example the set of reversed travel-time graphs of Fig. 3.14, in which the lower apparent velocities (see §2.3) have been labelled as "down-dip" apparent velocities V_{1d}, V_{2d} and V_{3d} (see equations 2.16 and 2.17). The branch V_{1d} should of course represent the direct wave and should therefore pass through the origin: it may however have a small intercept as it does in the figure, which must be due either to a thin superficial layer of velocity V_o, say, which is so thin that the direct wave through it is overtaken by the V_1 head wave before it reaches the first geophone, or to some error in the shot timing.

As long as the intercepts of the V_1 branches can be neglected, the velocities V_{1d} and V_{1u} should be equal: if they are not, the velocity of the surface layer is varying laterally and should if possible be investigated in more detail by shooting short spreads as indicated on Fig. 3.3.

If the three branches are produced to the end of the spread, then if the layering is planar the "end-to-end times" or "reciprocal times", T_{2d} and T_{3d}, should closely match the corresponding times, T_{2u} and T_{3u}, in the "up-dip" direction. If they do not, then either the allocation of points to branches is incorrect or the dips vary along the spread.

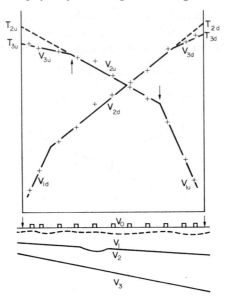

Fig. 3.14   Reversed refraction travel-time graphs of first arrivals over two dipping planar interfaces. An additional superficial layer might account for the small intercepts of the V_1 branches. Note the absence of true reversed coverage for the V_3 refractor.

The points should fit the branches to within the estimated reading error; those that do not either belong to an adjacent branch or indicate that the interface is significantly non-planar. For example, the point in Fig. 3.14 lying above the V_{2d} branch suggests a local deepening of the $V_1 - V_2$ interface, which is confirmed by the fact that the corresponding point on the V_{2u} branch is displaced by the same time.

Note that the two V_3 branches in this example show no "true reversal" at all, since they refer to separate parts of the $V_2 - V_3$ interface where its dip may be quite different, and that the V_2 branches give real reversed information only about that part of the spread between the two marked crossover ranges.

A complete treatment of a set of multilayer $t - x$ graphs for arbitrarily dipping interfaces may be found in (e.g.) Telford *et al.,* 1976, but we shall here consider only the two-layer case with a single interface dipping at an angle a as shown in Fig. 3.15(a). We first check that

$T_{2u} = T_{2d}$ within expected error limits, and then read from the graph the values of V_1, V_{2d}, V_{2u} and of t_{1d}, t_{1u}, the "intercept times" indicated in the figure. Then from equations (2.16) and (2.17) it follows that

$$\theta_c + a = \text{arc sin } V_1/V_{2d}$$

$$\theta_c - a = \text{arc sin } V_1/V_{2u}$$

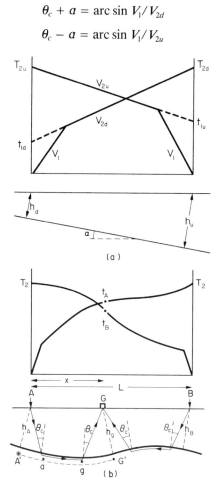

Fig. 3.15   (a) Reversed first-arrival travel-time graph for a single dipping interface. (b) Reversed first-arrival travel-time graph for a non-planar interface: the delay time at (for example) A is the time difference between the paths Aa (at velocity V_1) and $A'a$ (at velocity V_2).

Adding and subtracting these equations, we find that

$$\theta_c = (\text{arc sin } V_1/V_{2d} + \text{arc sin } V_1/V_{2u})/2 \qquad (3.2)$$

and
$$a = (\text{arc sin } V_1/V_{2d} - \text{arc sin } V_1/V_{2u})/2 \qquad (3.3)$$

Thus the dip a of the $V_1 - V_2$ interface is found directly, and the true velocity V_2 comes from θ_c by using

$$V_2 = V_1/\sin \theta_c \qquad (2.5)$$

The travel-time relationship for a dipping layer can readily be shown to be

$$t = x/V_{app} + 2h \cos \theta_c/V_1 \qquad (3.4)$$

which is conveniently of the same form as equation (2.7), with the slope of the branch given by $1/V_{app}$ and its intercept by $2h \cos \theta_c/V_1$ in which h must be taken to be the depth to the $V_1 - V_2$ interface measured *at the shotpoint* and *normal to the interface*. Thus for the two branches of Fig. 3.15(a) we can write

$$t_d = x/V_{2d} + 2h_d \cos \theta_c/V_1 = x/V_{2d} + t_{id} \qquad (3.5)$$

$$t_u = x/V_{2u} + 2h_u \cos \theta_c/V_1 = x/V_{2u} + t_{iu} \qquad (3.6)$$

in which h_d and h_u refer to depths measured as indicated above and on the figure.

Our interpretation can thus be completed by finding h_d from t_{id}, h_u from t_{iu}, and checking that the dip found from $\sin a = (h_u - h_d)/L$ (where L is the total spread length) agrees with that found directly from the apparent velocities as above.

Figure 3.15(b) shows how a reversed pair of $t - x$ graphs for a two-layer case might look if the $V_1 - V_2$ interface were far from planar, with the points deviating from straight-line branches by much more than the expected measurement errors. In this case a convenient method of interpretation named by Hagedoorn (1959) the "plus-minus" method can be applied to the points which record V_2 head-wave arrivals from both shotpoints. The head-wave travel-time equation (2.8) can be rewritten in the more general form

$$t = x/V_2 + t_{ds} + t_{dg} \qquad (3.7)$$

in which the intercept time t_i has been replaced by the sum of two "delay times" t_{ds} at the shot and t_{dg} at the geophone which for a horizontal boundary are equal to each other and to half the intercept time, i.e.

$$t_{ds} = t_{dg} = t_i/2 = h \cos \theta_c / V_1$$

The significance of the delay times when the boundary is irregular can be seen by comparing the ray path $AagG$ in Fig. 3.15(b) with a path $A'agG'$ supposed to be entirely in the V_2 layer. The time for the latter path is very nearly x/V_2 provided that the dip of the boundary does not exceed $10°$ or so, and so the delay times can be associated with the differences between the two paths. Since ag is common to both paths, the delay time t_d at A must be simply the difference between the travel times along the actual path Aa (at velocity V_1) and the hypothetical path $A'a$ at velocity V_2. Thus

$$t_{dA} = Aa/V_1 - A'a/V_2$$

$$= h_a/V_1 \cos \theta_c - h_A \tan \theta_c / V_2$$

$$t_{dA} = h_A \cos \theta_c / V_1 \tag{3.8}$$

using the same argument as that leading to the derivation of equation (2.7).

Obviously $t_{dG} = h_G \cos \theta_c / V_1$ and in general the delay time at any point is related to the depth (measured perpendicular to the interface as usual) at that point. Provided that the interface is nearly plane over the distance between G and the point of emergence of the corresponding ray from shotpoint B, the same delay time t_{dG} will appear in both the equations for the travel times from A and B to G. These are

$$t_A = x/V_2 + t_{dA} + t_{dG} \tag{3.9}$$

$$t_B = (L - x)/V_2 + t_{dB} + t_{dG} \tag{3.10}$$

since the range $BG = L - x$.

Adding these two equations gives

$$t_A + t_B = L/V_2 + t_{dA} + t_{dB} + 2t_G$$

$$= T_2 + 2t_G$$

where T_2 is the reciprocal time, that is, the head-wave travel time from A to B.

Thus
$$t_G = (t_A + t_B - T_2)/2 \qquad (3.11)$$

and the delay time at any geophone can thus be very simply found from the sum of the two observed travel times recorded there.

To convert t_G into the depth h_G we need the velocities V_1 and V_2: V_1 comes from the first branch of the $t - x$ graph as usual, but V_2 is best found by taking the difference of equations (3.9) and (3.10), giving

$$t_A - t_B = 2x/V_2 + t_{dA} - t_{dB} \qquad (3.12)$$

If the "minus time" $t_A - t_B$ is now plotted against range x, we obtain a graph which is much closer to a straight line than the original $t - x$ plots (since the variable delay time t_{dG} has been eliminated) and has a slope corresponding to a velocity of $V_2/2$. The intercept of this graph at $x = 0$ gives the difference of the two shotpoint delay times, and since their sum can be found from the reciprocal time

$$T_2 = L/V_2 + t_{dA} + t_{dB}$$

once V_2 is known, it follows that t_{dA} and t_{dB} can both be found.

In applying this method one must remember:

(a) to do so only to geophones recording headwaves from both shotpoints through the same refractor, i.e. those situated between the $V_1 - V_2$ crossovers;

(b) that it is applicable only when refractor dips are moderate;

(c) that the velocity V_1 is measured only outside the crossover points unless the branch is observed as a second arrival or additional short-range measurements are made. Using this value to find h_G from t_{dG} between the crossovers may not always be admissible.

3.53 *Complications in refraction interpretation: blind and hidden layers*

The methods of §3.52 or simple extensions of them will often suffice to give depths to the refractor of interest with a precision of the order of ± 10 per cent. There are however two important circumstances in which

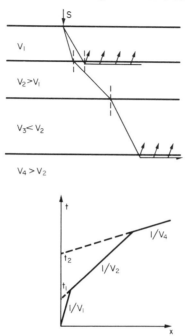

Fig. 3.16   The effect of a downward decrease in velocity.

the travel-time graph does *not* give complete information about ground layering, so that much greater errors can arise.

If one of the layers is of lower velocity than the one which overlies it, the rays entering it are refracted towards the vertical, and it is clear from Fig. 3.16 that no critically refracted ray, and therefore no head-wave returning to the surface, can exist. The branch of the travel-time graph corresponding to the layer of velocity V_3 is absent, and a head-wave is returned only when a layer of velocity greater than V_2 is reached. If the graph is the only evidence available, the intercept time t_2 will be used to find the depth to the top of this V_4 layer as though the V_3 layer had the higher velocity V_2, and an over-estimate of depth will result. The over-estimate may be considerable, being determined by the thickness and velocity of the low-velocity layer, and may be reduced if a neighbouring borehole indicates the probable existence of such a layer, though changes in its thickness and velocity between the borehole and the seismic line are in principle unknowable. A low-velocity layer may therefore be called a *blind layer*, though this term has also been used to describe the situation discussed below.

If one of a series of layers is thin in comparison with its depth, the head wave from it may never reach the surface as a first arrival, since the head wave from the layer below overtakes it at a range at which it still arrives later than the direct pulse. This situation can lead to a misinterpretation if *only* first arrivals are plotted on the graph. A good example of this is shown in Fig. 3.17, taken from a paper by Soske (1959). The first interpretation as a two-layer case (note that the velocity of 26.5 ft ms^{-1} or 7.9 m ms^{-1} is improbably high to represent a third layer, and so is more plausibly interpreted as an ''up-dip'' apparent velocity of the head wave from the 16 ft ms^{-1} refractor) is shown to be erroneous by a borehole near the shotpoint, which passes through a relatively thin bed of basalt of intermediate velocity. With the addition of the borehole control, or by recording the ''second arrivals'' of the head wave from the basalt, a correct interpretation is possible. Another example, on a much larger scale, of the possibility of such a *''hidden layer''* is the fact that the layer of sediment, perhaps a kilometre thick, on the bed of the oceans, does not contribute first arrivals to a refraction profile shot on the ocean surface: the depths and velocities involved in this case are in fact such that the sedimentary layer would have to be more than 8 km thick before it could be detected without making use of later arrivals. The greatest possible thickness of a hidden layer can readily be calculated from apparently ''two-layer'' data if a velocity is assumed for the intermediate layer (Green, 1962).

A refraction line on the scale of Soske's example or shorter, typical of work in engineering and groundwater geophysics, is very likely to give seismograms which are too complex for anything but first arrivals to be picked, so that thin intermediate layers remain ''hidden''. Data of higher resolution, however, such as are commonly obtained in larger scale seismology in which wavelengths are not so large compared with ranges and layer thicknesses, often show clear later arrivals so that a more complete interpretation can be made. Such an interpretation is still ''blind'' to low-velocity layers, though if the velocity varies continuously with depth as is common in the fields of crustal and global seismology, a

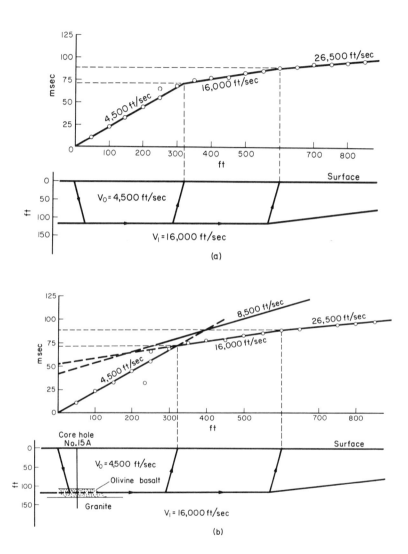

Fig. 3.17   The "hidden layer" in refraction seismology (from Soske, 1959). (a) Erroneous interpretation based on first arrivals only; (b) the structure shown by drilling; with calculated time-distance graph for head-wave from second layer.

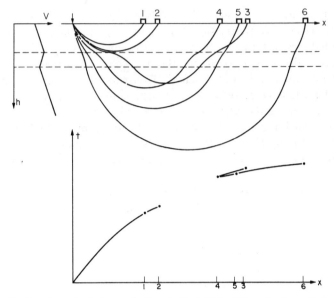

Fig. 3.18 Continuous velocity gradients including a "blind" layer lead to a shadow zone
with no refracted arrivals and a gap in the travel-time curve,

blind layer will lead to the existence of a "shadow zone" with a break in
the travel-time curve as indicated in Fig. 3.18. As the ray diagram shows,
no energy is returned to the surface from within the blind layer, and the
velocity distribution within it remains indeterminate even though its
presence may be deduced from the data.

Although ground models consisting of a few constant-velocity layers
separated by plane, dipping or gently undulating interfaces can explain
many of the travel-time graphs encountered in applied refraction
seismology, discontinuous lateral variations must also be considered.
Some examples are illustrated in Fig. 3.19, together with reversed travel-
time graphs which should be self-explanatory. Note that a decrease in
velocity from one travel-time branch to the next *must* be due to a lateral
change in velocity or to a downward dip of a refractor and *cannot*
indicate an underlying low-velocity layer, and that this feature, together
with the others illustrated, can be unambiguously interpreted if reversed
information is available.

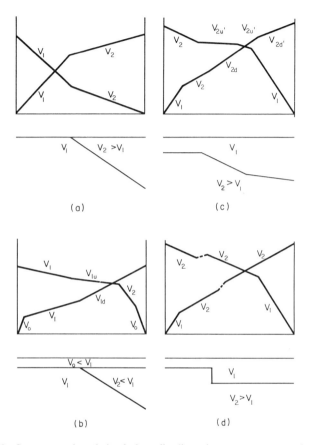

Fig. 3.19 Some examples of simple laterally discontinuous structures and schematic reversed refraction travel-time graphs that would be associated with them. (a) A lateral velocity change. The t–x graph is unchanged for any dip of the boundary so long as the higher velocity material overlies the lower. (b) If $V_2 < V_1$, branches of apparent velocities V_{1d}, V_{1u} are produced. The effect of an additional low-velocity surface layer is also shown. (c) An increase of refractor dip can also lead to a low velocity branch V_{2d} following one of higher velocity V_2. Note that a plane-layer interpretation is possible only if the branches V_{2u}–V_{2d} and V_{2u}–V_{2d} can be correctly paired. It will usually be easier to use the "plus-minus" approach (eqns. 3.11, 3.12). (d) The dipping segment of (c) is here steepened to a fault-like step. The steps in the t–x graphs are less sharp because of diffraction effects, and are offset by about $h \tan \theta_c$ from a step at depth h.

CHAPTER 4

Theory of Electrical Resistivity Surveying

4.1 Introduction

THOSE electrical methods in which current is applied by conduction to the ground through electrodes depend for their operation on the fact that any subsurface variation in conductivity alters the form of the current flow within the earth and this affects the distribution of electric potential, the degree to which it is affected depending on the size, shape, location and electrical resistivity of the subsurface layers or bodies. It is therefore possible to obtain information about the subsurface from potential measurements made at the surface. A sink or hollow in limestone filled with clay and concealed by a mantle of overburden is an example of a body of relatively good conductivity within a poorly conducting medium. There is a concentration of current flow through the clay fill in preference to the limestone, and a corresponding disturbance in the potential distribution in and around the sink, which can be measured at the surface if the sink is large enough or not too deeply buried.

The usual practice is to pass the current into the ground by means of two electrodes and to measure the potential drop between a second pair placed in line between these. If the ground is uniform, then from a knowledge of the potential drop, current and electrode spacings, using the appropriate expression, the ground resistivity can be calculated (electrical resistivity is the reciprocal of conductivity and is a measure of the resistance of a piece of the material of given size and shape – see equation (4.3)). Where the resistivity of the ground through which the current flows is not uniform the expression for uniform ground is still used but the quantity calculated is now called the "apparent resistivity".

As will be explained it is from an analysis of the variation of this quantity with changes in electrode position and spacing that deductions about the subsurface can be made.

4.2 Electrical Properties of Rocks

With the exception of clays and certain metallic ores the passage of electricity through rocks takes place by way of the groundwater contained in the pores and fissures, the rock matrix being non-conducting. All other factors being constant an increase in the concentration of dissolved salts in the groundwater leads to a decrease in resistivity. In a general way the resistivity is also controlled by the amount of water present. The more porous or fissured a rock the lower the resistivity. Degree of saturation also affects resistivity which increases with a decrease in the amount of water in the pore spaces and fissures.

When rocks are fractured and fissured it is not really practicable to quantify the relationships between resistivity, rock properties and electrolyte content. Rocks such as pure sandstones, in which the porosity is purely intergranular and in which the current is carried only by the electrolyte are more tractable. For example, in such rocks, where ρ is the resistivity when fully saturated with water of resistivity ρ_ω, the ratio $F = \rho/\rho_\omega$ for a particular formation tends to be constant and is known as the formation factor.

There is also a relationship between the formation factor F and the porosity ϕ having the general form

$$F = a/\phi^m \qquad (4.1)$$

where a and m are constants, their values being governed by the nature of the formation. Since m has a value not far from 2, the formation factor varies more or less inversely as the square of the porosity.

Unlike the other rock-forming minerals, which are in themselves non-conductors, conduction takes place through clays by way of the weakly bonded surface ions. Equation 4.1 above does not therefore apply to porous rocks containing any appreciable amount of clay minerals.

Metallic sulphides and some other metallic ores are semiconductors, with resistivities several orders of magnitude lower than the common rock-forming minerals, though still much higher than those typical of metals.

In crystalline rocks which have very low porosities conduction takes place mainly along cracks and fissures and, other factors being constant, the degree to which these are present will control the resistivity.

It will be realized from what has been said that the resistivities of formations range widely, not only from formation to formation but even within a particular deposit, this being particularly true of near surface unconsolidated materials. A sand, for example, can show variations in porosity and degree of saturation which may cause its resistivity to change considerably over quite a short distance. There is therefore no precise correlation of lithology with resistivity. Nevertheless, some generalization is possible, and the order of increasing resistivity for water-saturated rocks tends to be clay, sand and gravel, limestone, the values for crystalline rocks being higher still. In the literature widely different values are given for the resistivities of each particular rock type, this being due not only to the variation of lithology within a particular group, but also to the conditions of saturation and water resistivity under which the resistivity was measured. The values given in Table 4.1 give the range of resistivities usually encountered for some common rock types but they do not represent extreme limits.

Table 4.1   The electrical resistivity of rocks. The resistivities of rocks are strongly influenced by fissuring, porosity, groundwater conductivity and saturation so that only approximate ranges can be given.

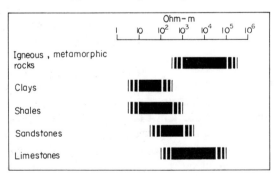

4.3 Theory of Current Flow in a Homogeneous Medium

For the purpose of quantitative interpretation the ground is regarded as consisting of regions of approximately constant resistivity separated from others of differing resistivity by plane interfaces. These interfaces correspond generally to boundaries between strata of different lithology, or to faults. For a proper appreciation of the problem of interpretation it is necessary therefore to understand the behaviour of current flow in layered media and how this affects the distribution of potential. In dealing with such problems the starting point is Ohm's Law

$$\frac{V}{I} = R \qquad (4.2)$$

where I = current in a conducting body,

V = potential difference between two surfaces of constant potential,

R = constant called the resistance between the surfaces.

We must also introduce at this point the definition of resistivity. If a conductor carries a current with parallel lines of flow over a cross-sectional area A, then its resistivity ρ is defined by

$$\rho = \frac{RA}{L} \qquad (4.3)$$

where R is the resistance measured between two equipotential surfaces separated by a distance L (see Fig. 4.1). It follows from this equation and the previous one that the total current over the area A is

$$I = \frac{V}{R} = \frac{VA}{\rho L}$$

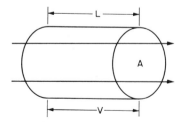

Fig. 4.1   The definition of resistivity.

and that the "current density" j is given by

$$j = \frac{I}{A} = \frac{V}{\rho L} \tag{4.4}$$

If the lines of current flow are not parallel, so that the current density varies over the conductor, this same argument can be applied to an infinitesimal element of the conductor bounded by equipotential surfaces which may be curved. The ratio V/L becomes in the limit the potential gradient dV/dL, and the expression of Ohm's Law is the equation

$$j = -\frac{1}{\rho} \cdot \frac{dV}{dL}$$

The negative sign has been introduced here to express the fact that potential increases in the *opposite* direction to the current flow. The component of the current density in a direction r is

$$j_r = -\frac{1}{\rho} \cdot \frac{\partial V}{\partial r} \tag{4.5}$$

in which the potential gradient in the direction r is used thus being always less than that in the direction of the current flow. Note the important fact that in a homogeneous medium an increase in current density, seen as a crowding or convergence of current lines, means an increase in the magnitude of the potential gradient. Conversely a divergence means a decrease in gradient.

The next step in the development of the theory should be to derive the potential in a homogeneous medium due to a point source of current (clearly for current to flow there must be a positive "source" and at some other point a negative "sink" as it is called, but here and in the following sections, unless otherwise stated, the "sink" is assumed to be so far away that its effect can be neglected). The potential due to the source varies inversely with distance and is therefore given by

$$V = \frac{S}{r} \tag{4.6}$$

where S is a constant which can be considered to be the strength of the source and r is the distance from it (note the similarity of the expression for the potentials of a point mass and a magnetic pole). The value of S depends on the resistivity, the current and the situation of the source. Firstly imagine the medium to be infinite in extent. Consider the current which flows outward through a sphere of radius r surrounding the source. The current flowing through unit area of the surface of this is (from 4.5 and differentiating 4.6)

$$j = -\frac{1}{\rho} \cdot \frac{\partial V}{\partial r} = S/\rho r^2 \qquad (4.7)$$

Since the current density is the same over the whole spherical surface of area $4\pi r^2$ the total current is

$$I = 4\pi r^2 \cdot \frac{S}{\rho r^2} \qquad (4.8)$$

Therefore

$$I = 4\pi S/\rho \qquad (4.9)$$

That is,

$$S = I\rho/4\pi \qquad (4.10)$$

However, if the medium is only semi-infinite and is bounded by a plane surface separating it from the air, which can be taken to be of infinite resistivity, and the source is located on the interface, a different result is obtained. This is of course the condition that is met in practice in resistivity surveying. Here again

$$V = S/r \qquad (4.6)$$

but under these conditions the current flows out through a hemisphere only and

$$S = I\rho/2\pi \qquad (4.11)$$

The strength of the source is then, as one might expect, twice the previous value.

If the potential gradient be measured at some point distant r from the source at the surface, then from this and a knowledge of the current strength the resistivity can be found, for (from 4.6 and 4.11)

$$V = I\rho/2\pi r$$

$$\therefore \frac{\partial V}{\partial r} = -I\rho/2\pi r^2$$

$$\therefore \rho = -\frac{2\pi r^2}{I} \cdot \frac{\partial V}{\partial r} \tag{4.12}$$

($\partial V/\partial r$ has the opposite sign to I, so that ρ is of course positive). In practice an approximate value of the potential gradient at any point can be obtained from a measurement of the potential difference between two closely spaced electrodes symmetrically placed with respect to the point and in line with the source. If ΔV is the potential difference between the two electrodes and ΔL their distance apart the potential gradient is $\Delta V/\Delta L$.

4.4 Two Media of Differing Resistivity

Figures 4.2(a) and 4.2(b) show theoretical cases of point sources of current, each embedded in a homogeneous medium and placed near a plane boundary across which there is a change of resistivity. In Fig. 4.2(a) the source lies in the lower resistivity medium (ρ_1). In the second figure, however, it is placed in the higher resistivity medium (ρ_2). In both instances lines of current flow have been drawn.

In a uniform medium well away from any resistivity boundaries current lines diverge radially from a source. Near a boundary this simple pattern is distorted. In Fig. 4.2(a) the current lines can be seen to diverge abnormally from the radial pattern as if seeking to avoid the higher resistivity medium. Where the source is in the higher resistivity medium, as shown in Fig. 4.2(b), there is, on the contrary, less divergence of current lines as the boundary with the lower resistivity medium is approached. It can be shown that at the boundary the current lines are refracted according to the rule

$$\frac{\tan \theta_1}{\tan \theta_2} = \frac{\rho_2}{\rho_1} \tag{4.13}$$

where θ_1 and θ_2 are the angles the current lines make with the normals to the boundary.

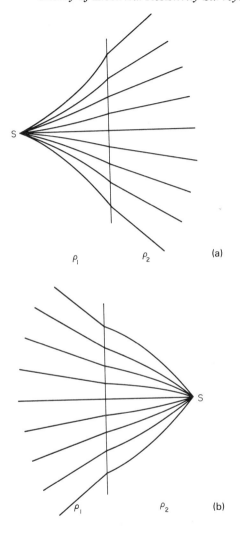

Fig. 4.2   The effect of a boundary on the distribution of lines of current flow for $\rho_2 > \rho_1$.
In (a) the source is in ρ_1 and in (b) in ρ_2.

Across the boundary from the source the current lines again become straight and appear to diverge radially from the source position.

Abnormal crowding of current lines indicates an increased potential gradient. Where the divergence is greater than in the undisturbed case the potential gradient is less than normal.

We note at this point that the pattern of current lines is not affected if the source is at the earth's surface (except that of course no current flows in the air) and the boundary is, for example, a vertical fault between rocks of differing resistivities.

Using the formula

$$\rho_a = -\frac{2\pi r^2}{I} \cdot \frac{\partial V}{\partial r} \tag{4.12}$$

we note also, therefore, that the apparent resistivity measured near a boundary will not be the true resistivity of the medium. As the boundary is approached the resistivity will rise or fall from the true value in the medium depending from which side the boundary is being approached.

The calculation of potential differences and apparent resistivities near a vertical plane boundary for a surface source may be approached quantitatively in a simple way using image theory. Source and boundary are shown in plan in Fig. 4.3(a). By analogy with optics the boundary may be thought of as partly "electrically reflecting". The potential distribution on the source side of the boundary may be calculated by making use of the fact that the effect of the boundary on the potential and therefore on the current pattern is equivalent to that of a second source placed at the "image point". The positions of source and image are shown in Fig. 4.3(b). The image Q is at a distance "behind" the boundary equal to that of the source. Its strength is kS where S is the source strength and k a constant depending on the resistivity values. The value of k lies between ± 1 and the image strength is less than that of the source except where ρ_2 is infinite in which case the boundary is totally "reflecting". We are, however, by supposing an image to be present, at the same time assuming that the boundary no longer exists, and that the medium containing the source extends everywhere (except above the ground surface). The potential at any point M on the surface in the lower resistivity medium as shown in Fig. 4.3(b) is therefore given by

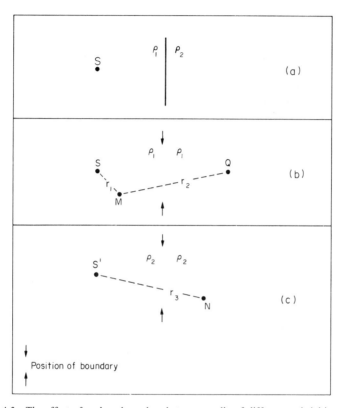

Fig. 4.3   The effect of a plane boundary between media of different resistivities on the potential. The diagram is a plan view. (a) A surface source S is shown near a vertical boundary between media of resistivities ρ_1 and ρ_2 where $\rho_2 > \rho_1$. (b) To calculate the potential at M by image theory the boundary is replaced by the image Q, the medium of resistivity ρ_1 now being supposed to extend over the full half space. (c) To calculate the potential at N the source is replaced by a source $S'(1 - k)$, ρ_2 now extending over the full half space.

$$V_1 = \frac{S}{r_1} + \frac{Sk}{r_2}$$

$$S = \frac{I\rho_1}{2\pi}$$

$$V_1 = \frac{I\rho_1}{2\pi}\left(\frac{1}{r_1} + \frac{k}{r_2}\right) \tag{4.14}$$

Looking back from any point N on the surface in the medium of resistivity ρ_2 to the source (Fig. 4.3(c)) and once again removing the boundary and permitting ρ_2 to extend everywhere we found that to calculate the potential correctly at N we now have to modify the strength of the source to S' where $S' = I\rho_2(1-k)/2\pi$. (Since the boundary "reflects" back a fraction k of the strength of the source, as seen from the higher resistivity medium the source has a strength S' $(1 - k)$, where in this instance, since we are removing the boundary and allowing ρ_2 to extend everywhere, $S' = I\rho_2/2\pi$.)

The potential in ρ_2 is therefore

$$V_2 = I\rho_2(1 - k)/2\pi r_3 \tag{4.15}$$

By considering a point on the boundary across which there can be no change in potential and thus equating V_1 and V_2 the *reflection coefficient* k is found to have the value

$$k = \frac{\rho_2 - \rho_1}{\rho_2 + \rho_1} \tag{4.16}$$

From a knowledge of the potentials the potential gradient is easily determined and thus the apparent resistivity can be calculated using

$$\rho_a = -\frac{2\pi r^2}{I}\cdot\frac{\partial V}{\partial r} \tag{4.12}$$

Note now that if the electrode separations are kept constant ρ_a is merely a measure of how the potential gradient and (since $\partial V/\partial r = -j\rho$) the current density at the surface vary as the array is moved from place to place.

We are now better able to understand the behaviour of current flow near a boundary. Where the source is situated in the lower resistivity medium the image across the boundary has the same sign as that of the source, since k is positive. The potential gradient due to the image thus opposes that of the source, leading to a reduced resultant gradient. Hence the current lines show a divergence from their normal radial pattern round the source. When the source is in the higher resistivity medium the image has the opposite sign and the two gradients add, thus increasing the potential gradient and producing a crowding or convergence of current lines.

We are also able to explain the various sections of a resistivity plot as a constant separation array consisting of a current electrode and two closely spaced potential electrodes is traversed across a fault from a lower to a higher resistivity medium. Such a plot is shown in Fig. 4.4.

As the potential electrodes approach the boundary the current density falls and with it the calculated value of ρ_a to a value less than ρ_1.

When they cross it the true resistivity increases from ρ_1 to ρ_2 and the apparent resistivity also increases discontinuously by a factor of ρ_2/ρ_1 to a value of $2\rho_1\rho_2/(\rho_1+\rho_2)$. Whilst the source and potential electrodes are on opposite sides of the boundary this constant intermediate value of ρ_a is recorded. Finally, when the current electrode enters the high resistivity medium, the lines of current flowing from it tend to converge towards the boundary. The current density in the region of the potential electrodes is thus reduced and the apparent resistivity recorded is thus still less than the true resistivity ρ_2. As the array is moved away from the boundary its influence decreases and the apparent resistivity gradually rises to its true value of ρ_2.

When a resistivity boundary is horizontal (see Fig. 4.8) and parallel to the surface, as in layered strata, the problem is complicated by the fact that there are not two layers 1 and 2 but three, the air (medium 0) being the third. In these circumstances an infinite number of images is produced of a source placed at the surface. An image Q_{+1} of S_+ is formed by the 1–2 boundary, at a distance $2h$ below the source where h is the thickness of medium 1. An image of Q_{+1} itself appears at Q'_{+1}, distance $2h$ above S_+, due to the presence of the 1–0 interface. This in its turn is further reflected by the 1–2 plane, appearing at Q_{+2}, distance $4h$ below S_+, and so on, the strength of the images decreasing with successive "reflections". The resulting potential at, say, a point on the surface is the sum of the potentials due to the source and its infinite series of images. This sum converges and so the potential can be calculated.

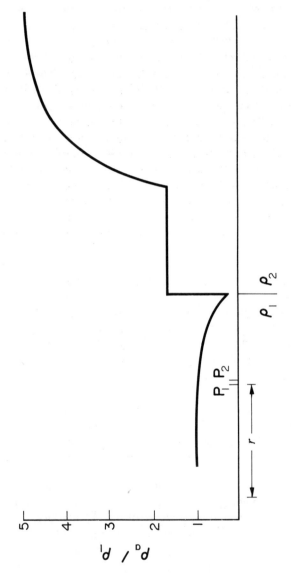

Fig. 4.4   The variation of apparent resistivity across a vertical boundary measured with a constant separation array of three electrodes.

4.5 The Four Electrode System of Measurement

All resistivity measurements could be made using a single current electrode and a closely spaced pair for measuring potential gradient as we have already described. However, it is not always practicable to have a fourth fixed and distant current electrode and there are also often good scientific reasons for using some symmetrical arrangement of four moving electrodes. One way of constructing a four electrode array would be to bring in the distant current electrode of the gradient array and place it so as to form a collinear configuration with the potential pair spaced symmetrically between the two current electrodes. Since the potential at any point is the sum of the positive potential due to the source and the negative potential due to the sink, the resulting potential gradient will also be the sum (taking account of the direction, that is, the *vector* sum) of the gradients due to the two electrodes separately. In practice an array of this type, known as a Schlumberger array, uses a ratio of current to potential electrode separation of about 10.

A somewhat similar configuration, much used in Britain, is the Wenner array. In this arrangement the four electrodes are placed in line at equal distances apart, the inner two again being normally used as the potential electrodes, though it is sometimes convenient to use one of the other two possible pairs. With a finite separation of the potential electrodes it is of course still possible to use equation (4.14) to derive the apparent resistivity from the measured potential difference, which is simply the integral of the gradient over this separation. A more direct approach is based directly on equations (4.6) and (4.11) for a homogeneous medium, from which the potential at each electrode, and hence the potential difference between them, can quickly be found as described below. If the distances of the potential electrodes from the current electrodes are as shown in Fig. 4.5(a) then the potentials at P_1 and P_2 are

$$V_1 = \frac{I\rho}{2\pi}\left(\frac{1}{r_1} - \frac{1}{r_2}\right)$$

$$V_2 = \frac{I\rho}{2\pi}\left(\frac{1}{R_1} - \frac{1}{R_2}\right)$$

The potential difference between P_1 and P_2,

$$V_1 - V_2 = \Delta V = \frac{I\rho}{2\pi} \left(\frac{1}{r_1} - \frac{1}{r_2} - \frac{1}{R_1} + \frac{1}{R_2} \right)$$

Hence the resistivity

$$\rho = 2\pi \, \Delta V/I \left(\frac{1}{r_1} - \frac{1}{r_2} - \frac{1}{R_1} + \frac{1}{R_2} \right) \qquad (4.17)$$

If the electrode separations are made the same and equal to a then

$$r_1 = R_2 = a \quad \text{and} \quad r_2 = R_1 = 2a.$$

Hence, substituting in equation (4.17),

$$\rho = 2\pi a \, \Delta V/I \qquad (4.18)$$

This, then, is the appropriate expression when a Wenner spread, shown in Fig. 4.5(b), is used. Where the ground is inhomogeneous, as previously explained the quantity calculated is described as the apparent resistivity, ρ_a.

When the Schlumberger arrangement (Fig. 4.5(c)) is used the apparent resistivity may be calculated from equation (4.17) putting in appropriate values for the electrode separations. Alternatively it may be calculated from a simpler but approximate expression which is derived from this equation by putting $r_1 = R_2 = r$ and $R_1 = r_2 = r + \Delta r$.

Thus

$$\rho = \frac{2\pi \, \Delta V}{I} \cdot \frac{1}{2\left(\dfrac{1}{r} - \dfrac{1}{r + \Delta r} \right)}$$

$$\approx \frac{\pi r^2}{I} \cdot \frac{\Delta V}{\Delta r} \qquad (4.19)$$

Note that this is identical with equation (4.12) except for the absence of a factor 2. This is because the two current electrodes in the Schlumberger configuration are equidistant from the potential electrodes, so that the potential difference is twice that observed when one current electrode is at infinity.

An exact form of equation (4.19) can be obtained by setting $r = n.\Delta r$ when we have

$$\rho_a = \pi \frac{\Delta V}{I} . n(n+1) \Delta r$$

Another arrangement is the Lee Partition spread (Fig. 4.5(d)) which is in essence a Wenner arrangement with a third potential electrode placed midway between the other two. This enables two measurements to be made, one for the left and one for the right half of the spread, without the necessity of moving the current electrodes. The apparent resistivity is given by

$$\rho_a = 4\pi a . \frac{\Delta V}{I} \qquad (4.20)$$

where ΔV = potential difference between the centre and one outer potential electrode. If the two measured values of V are identical, it can be taken that the ground is laterally homogeneous, as is assumed during interpretation.

Less commonly used for resistivity work but important for some special purposes and for Induced Polarization surveys (q.v.) are dipole-dipole arrays (Fig. 4.5(e)). These consist of two pairs of equally spaced electrodes, a current pair and a potential difference measuring pair. The distance between the pairs is normally greater than their individual spacing. Various arrangements are possible depending on the relative orientation of the pairs. The simplest is a collinear system of the form CCPP for which the apparent resistivity may be calculated from

$$\rho_a = \pi \frac{\Delta V}{I} n . (n+1)(n+2)a \qquad (4.21)$$

A special variant is one in which the electrodes are placed at the corners of a square (Fig. 4.5(f)), making it possible to average out azimuthal variations of resistivity such as may be found over contacts or steeply dipping beds (Habberjam and Watkins, 1967). The apparent resistivity is given by

$$a = \frac{2\pi \Delta V}{I} \cdot \frac{a}{(2-\sqrt{2})} \qquad (4.22)$$

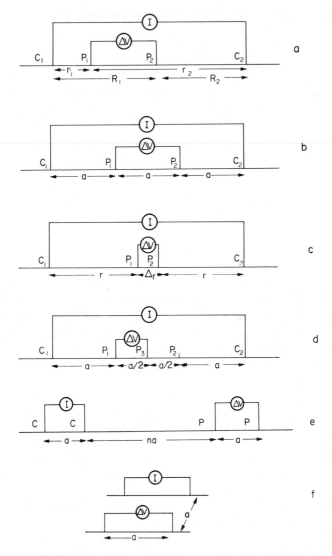

Fig. 4.5   (a) The four electrode array. (b) Wenner configuration. (c) Schlumberger configuration. (d) Lee Partition configuration. (e) Dipole–dipole configuration. (f) Square configuration.

4.6 Measurement with an "Expanding" Electrode System

In Fig. 4.4 an illustration has already been given of how a vertical boundary may be located by horizontally traversing a three electrode gradient array with fixed separation across the contact.

Very often what is of interest is the variation of resistivity in the vertical direction. Either structure at depth is to be located, or more often the ground is horizontally layered, and it is the layering that is to be determined.

To investigate variation of resistivity with depth a set of measurements is made at one station with gradually increasing electrode separations. With the three electrode gradient array for example, at each measurement the spacing between the current electrode and the gradient measuring pair is increased. If the Schlumberger array is used, then both current electrodes are expanded outwards from the potential pair, the spacing of which is not increased until the potential difference becomes too small to measure satisfactorily. When the Wenner configuration is expanded the centre point of the array is kept fixed and the distance *a* between electrodes increased at each measurement by a factor determined by the resolution required from the survey. Whichever array is used the data are presented in the form of a "resistivity curve" of measured apparent resistivity plotted against electrode separation with both variables usually on a logarithmic scale.

It is often said that as the separation of current electrodes is increased current penetration into the ground increases, leading to a greater "depth of investigation". This is not strictly correct. Consider, for example, the simple example of a thin horizontal high resistivity layer in an otherwise homogeneous medium, under investigation using a single current source (the sink being very distant) and two closely spaced potential electrodes. All the information needed to calculate the resistivity variation with depth can be obtained by measuring the potential gradient out from the single current electrode. In this instance there is clearly no change in the current distribution in the ground between successive measurements, for the current electrode is not moved. However, analysis shows that the disturbance to the horizontal potential gradient at the surface due to the thin high resistivity layer increases outwards from zero at the source to a maximum and then decreases again.

Thus, in plotting an apparent resistivity curve derived from the gradient measurements, the peak of the disturbance in the apparent resistivity will be roughly related to the layer depth. The depth relationship is by no means exact, for it depends also on resistivity contrast and layer thickness. Bringing in the second current electrode of opposite polarity to a finite distance does not fundamentally affect the situation. The horizontal surface gradients add, the nearest current electrode to the measuring pair having the most weight.

Roy and Apparao (1971) have calculated "depths of investigation" for various types of arrays. Using an electrostatic analogy, they suppose that the potential at the surface is a sum of contributions from electrically polarized volume elements of the medium. They calculate the contributions to the measured potential difference at the surface from a thin horizontal layer of the medium as its depth is increased and show that this passes through a maximum, which they define as the depth of investigation of the array. The model is of course an unrealistic one, for we are normally interested in a stratified or inhomogeneous medium. Even so, except where contrasts are high it can be a useful guide to the appropriate range of electrode separations required for exploring a particular situation. For a Wenner array the depth of investigation is about one-third the distance a between the electrodes, so that the overall length of such an array needs to be about ten times the maximum depth of interest.

4.7 Resistivity Curves for Horizontally Layered Strata and their Interpretation

In the field graphical methods involving curve matching are usually used to obtain approximate solutions to resistivity curves for horizontally layered strata. In some circumstances the results obtained are acceptable but it is normal practice when a computer is available to calculate a resistivity curve from the layer thicknesses and resistivities obtained graphically and then to compare the computed and observed curves. If the fit is not close the thickness and resistivity values are modified and the curve recomputed, the process being repeated until the fit is acceptable. This is therefore a trial and error procedure, an indirect method of interpretation. It is still a much used approach because the computation of resistivity curves using given values for resistivities and layer thicknesses is quick and comparatively simple. The direct interpretation of an observed resistivity curve in terms of layer thicknesses and resistivities is more complex and computational procedures for doing this have only recently been developed (Zohdy, 1975; Marsden, 1973). They are likely to become standard in the near future but a good deal of development work still remains to be done on the various alternative approaches to the problem.

It will be helpful to start discussion of interpretation by considering the form of the resistivity curve for some layered situations. Take firstly the example of a single horizontal layer overlying a substratum of differing resistivity and infinite thickness. When the electrode separation is small compared with the thickness of the top layer the measured potential difference is hardly influenced by the presence of the boundary since its distance is large compared with the separation. The measured resistivity thus approximates to that of the top layer. As the electrode separation is increased the depth of investigation increases and with it the influence of the lower layer. Hence at very large separations the measured apparent resistivity approximates to that of the lower layer. Figure 4.6 shows, for a

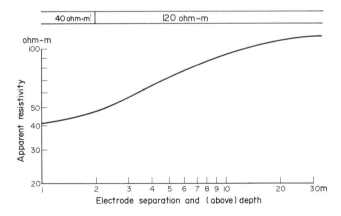

Fig. 4.6   The variation of apparent resistivity with electrical separation for a two layer earth where $\rho_2 > \rho_1$.

Wenner array, a typical "two layer" curve in which apparent resistivity is plotted against electrode separation. If (as in the diagram) a double logarithmic scale is used the *shape* of the curve depends only on the resistivity contrast and its *position,* relative to the two axes, on the top layer resistivity and the layer thickness. Using a double logarithmic scale thus simplifies presentation, and as will be explained later, greatly helps in interpretation.

Where the layering is simple qualitative interpretation by inspection is possible. The shape of the curve in Fig. 4.6 suggests that only two layers are present, and some estimate of layer resistivities is possible from the limiting values at small and large spacing. With experience a guess may be made as to the thickness of the top layer. Three layer curves can be recognized easily when the middle layer has a higher or lower resistivity than the other two, provided the contrast is not too small or the layer too thin. If the second layer has an intermediate resistivity it may be difficult to differentiate the resistivity curve from a two layer type. Examples are given in Fig. 4.7.

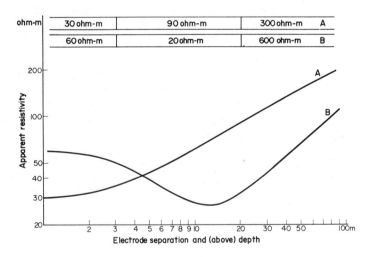

Fig. 4.7   The variation of apparent resistivity with electrode separation for a three layer earth where (A) $\rho_1 < \rho_2 < \rho_3$ and (B) $\rho_2 < \rho_1 < \rho_3$.

Given the apparent resistivity curve of any set of horizontal layers of infinite extent and differing resistivities it is theoretically possible to derive a unique solution in terms of layer thicknesses and resistivities. In practice this is by no means always so. Even when approximating to a horizontally layered medium the ground always departs to some extent from the ideal model. As a consequence the shape of the resistivity curve is affected and ambiguities are introduced which may result in an incorrect interpretation or, where there is gross departure of the ground from a layered form, make interpretation impossible. This and the more fundamental factors of equivalence and suppression, discussed later, impose very considerable limitations on the method so that interpretation tends to be confined to solutions involving only a few layers. Such a limitation is perhaps not so serious in practice as it may sound.

For simplicity and illustrative purposes we shall start quantitative treatment of interpretation procedures by employing image theory to calculate two layer apparent resistivity curves.

The ground conditions assumed are those described in §4.4 but instead of a single source there are now, as shown in Fig. 4.8, both a positive source and a negative sink, the two being of opposite sign and a finite distance apart. The source and the sink are, of course, the two electrodes by means of which current is passed through the ground. The combined effect of the ground to air interface and the 1–2 interface is to produce infinite series of images of both the source and of the sink, S_+ and S_-, the potential at any point on the earth's surface being merely the sum of the potentials due to the source, the sink, and their respective images. In terms of current, formation resistivities, electrode separation and interface depth the potential difference ΔV between the inner electrodes of a Wenner spread can be written

$$\Delta V = \frac{\rho_1 I}{2\pi} \left[1 + 4 \sum_{n=1}^{\infty} \frac{k^n}{[1 + (2nh/a)^2]^{1/2}} - 2 \sum_{n=1}^{\infty} \frac{k^n}{[1 + (nh/a)^2]^{1/2}} \right] \qquad (4.23)$$

i.e. $$\Delta V = \frac{I\rho_1}{2\pi a}(1 + 4F) \qquad (4.24)$$

in which F is a sum of terms involving the depth of the interface, the coefficient k and the electrode spacing. From this

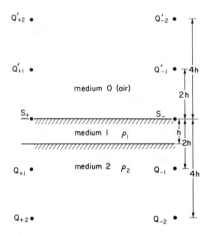

Fig. 4.8   The position of the nearer images due to a source and a sink for a two layer medium.

$$\frac{2\pi a \Delta V}{I} = \rho_1(1 + 4F) \qquad (4.25)$$

The left hand side of this equation is the formula for the apparent resistivity when using a Wenner electrode system. We can write therefore

$$\rho_a = \rho_1(1 + 4F)$$

or

$$\rho_a/\rho_1 = 1 + 4F \qquad (4.26)$$

The equation is in a dimensionless form. Figure 4.9 shows the set of two layer curves calculated from the equation, ρ_a/ρ_1 being plotted against h/a for the full range of k. A double logarithmic scale is used so we can write $\log \rho_a - \log \rho_1 = \log (1 + 4F)$ and $\log (a/h) = \log a - \log h$. For any given curve the coordinates of a point are $\log (\rho_a/\rho_1)$, $\log (a/h)$, i.e. $(\log \rho_a - \log \rho_1)$ and $(\log a - \log h)$. Provided k is kept constant the effect of multiplying either ρ_1 or h by some factor is to translate the curve unmodified along the appropriate axis since multiplication merely increases the value of $\log \rho_1$ or $\log h$. If ρ_1 is changed but k kept constant then ρ_2 must at the same time alter accordingly, since $k = (\rho_2 - \rho_1)/(\rho_2 + \rho_1)$.

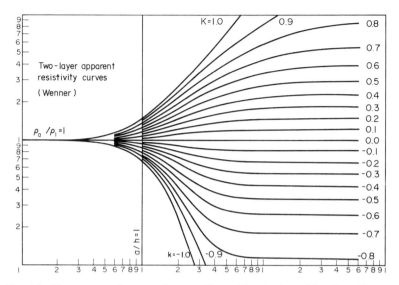

Fig. 4.9   Master curves for a two layer medium with a horizontal interface (Wenner configuration).

To carry out an interpretation the field resistivity curve is plotted on transparent paper laid over double logarithmic paper with the same modulus as that used for the type curves. If the field resistivity curve is a two layer curve it must have the same form as one of the type curves or be of a form that can be interpolated between two adjacent ones. The field curve is therefore superimposed on the appropriate set of type curves and, as shown in Fig. 4.10, keeping the two sets of axes parallel, it is moved about until a fit is obtained, or until it appears to be interpolated between adjacent type curves in the correct place. Now the $\rho_a/\rho_1 = 1$ axis of the master curves cuts the ordinate of the field curve at $\rho_a = \rho_1$. Similarly the $a/h = 1$ axis of the type curves must cut the abscissa of the field curve where $a = h$. The reflection coefficient for the field curve is that of the type curve to which it is the closest fit, or if the curve is interpolated it is intermediate between that of the nearest type curves on either hand.

The curve fitting method can be extended directly to three and four layer problems. A set of 2400 theoretical Wenner type 3 and 4 layer

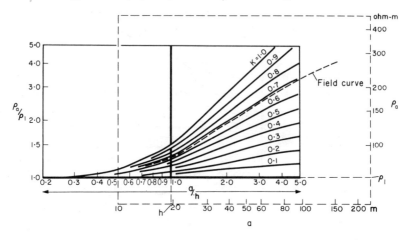

Fig. 4.10   The interpretation of a two layer resistivity curve by comparison with the master set. The top layer resistivity $\rho_1 = 68$ ohm-m and the depth to the interface $h = 19.5$ m.

curves has been published by Mooney and Wetzel (1956) and a smaller selection of 3 layer curves for use with the Schlumberger array by the Compagnie Générale de Géophysique (1955).

The procedure for interpretation is virtually the same as that used in dealing with two layer structures. However, even the large number of published curves is quite inadequate and does not cover anything like the range of conditions met in practice. Of the various techniques that have been evolved to meet this difficulty "partial curve matching" is perhaps the most used. The method is most simply illustrated with reference to three layer resistivity curves, but it can be used to interpret curves for four or even more layers. As an example a curve such as that shown in Fig. 4.11 is taken in which the second layer has a low resistivity, the third layer resistivity being high. For small values of electrode separation the shape of the curve will be unaffected by the presence of the third layer and so it approximates to a two layer curve with h_1 the thickness of the upper layer and $k = (\rho_2 - \rho_1)/(\rho_2 + \rho_1)$. Interpretation can therefore be carried out using a set of standard two layer type curves. The lower part of the curve also approximates to a two layer type for which (in this particular instance) the upper layer has a thickness equal to the sum of the two upper layers and a resistivity given by the equation

$$\frac{h_r}{\rho_r} = \frac{h_1}{\rho_1} + \frac{h_2}{\rho_2}$$ (4.27)

ρ_r being known as the "replacement resistivity", and $h_r = h_1 + h_2$ as the replacement thickness. Since $\rho = 1/\sigma$ (where σ is conductivity) we can equally write $(h_1 + h_2)\sigma_r = h_1\sigma_1 + h_2\sigma_2$. The term $h\sigma$ is the conductance of a layer, and the expression states that the conductance of the replacement layer is the sum of the conductances of the individual layers. In the example being considered, at large separations current flow is more or less parallel to the layers and the expression is approximately true. The third layer of the curve, with resistivity ρ_3, is of course the bottom layer and k_2 the contrast between the replacement layer and the third layer has the value $(\rho_r - \rho_3)/(\rho_r + \rho_3)$.

Where the resistivity sequence of the layers is different from that of the example the current will not flow parallel to the layers and different

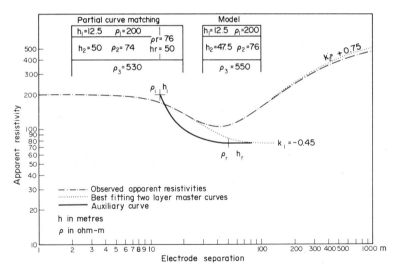

Fig. 4.11   Interpretation of an apparent resistivity curve for an earth with three layers by the partial curve matching method.

expressions have to be used to calculate replacement resistivity and thickness, the latter in general not being equal to the sum of the two top layers. The thicker the second layer the better the lower section of the curve approximates to one of a two layer type. Even so there is usually some ambiguity about the fit to this lower part of the curve. To reduce this use is made of auxiliary curves.

An auxiliary curve is the locus of the point ρ_r, h_r as h_2 varies. For $h_2 = 0$, $\rho_r = \rho_1$ and $h_r = h_1$. After fitting a two layer curve to the first part of the three layer curve and obtaining values for ρ_1, h_1 and k (and therefore ρ_2) the appropriate auxiliary curve is drawn on the observed resistivity curve plot with its origin at $\rho_1 h_1$. The set of 2 layer type curves is now used to fit the lower part of the curve, the origin of the type curves being constrained to lie on the auxiliary curve, as it must, since we are fitting a two layer curve with an upper layer of resistivity ρ_r and thickness h_r. The process is illustrated in Fig. 4.11. Having obtained a fit the origin of the two layer curve now gives ρ_r and h_r and $k_2 = (\rho_3 - \rho_r)/(\rho_3 + \rho_r)$. A simple graphical procedure, which will not be described here, now enables h_2 to be found from h_r. If more than three layers are present successive segments of the resistivity curve can be fitted to two layer curves, a new replacement resistivity and thickness being found at each stage, from which the true layering can be calculated.

Sets of auxiliary curves for various conditions have been calculated. In particular Koefoed (1960) has shown that the whole range of auxiliary curves can be combined into a single diagram accurate enough to be of general use.

Once having obtained an approximate but acceptable solution by curve matching the layer parameters can be used in a curve generating computer programme to test agreement between field and computed resistivity curves. If this is not satisfactory resistivities and thicknesses are modified until a good fit is secured.

Image theory is too clumsy to be used for the calculation of multilayer resistivity curves. Resistivities are instead calculated from the fundamental expression for the potential in a layered medium, first derived by Stefanescu. A very elegant and quick method of computing resistivity curves based on this expression has been given by Ghosh (1971).

Many factors affect the accuracy of an interpretation and it is often not possible to say with any certainty in a particular situation which are important. The degree to which the ground approximates to the model of uniform layers is one factor already mentioned. More fundamental ambiguities may also arise. It is by no means uncommon for a range of solutions to be found which fit the apparent resistivity curve equally well within acceptable limits of error, there being very little or hardly any appreciable difference in the calculated curve shapes. Such a range of solutions is obtained by making interdependent changes in the thickness and resistivity of a single bed. Their existence is most simply exemplified by a three layer earth with a low resistivity middle layer through which the current may be considered to flow mainly horizontally. If the resistivity and thickness are altered together over quite a wide range, but in such a way that the conductance $S = h\sigma$ is kept constant, then the shape of the resulting resistivity curve will change very little. In extreme circumstances the thickness or resistivity may be changed 100 per cent or more with a corresponding change in the resistivity curve of less than 5 per cent. If this is the accuracy to which the resistivity curve is known, then the whole range of solutions is equally valid and is said to show "equivalence".

In the situation in which a high resistivity bed occurs in a sequence immediately above a low resistivity layer current tends to flow across the bed and the parameter that must be kept constant for equivalence to hold is $T = h\rho$, the transverse resistance. Unfortunately uncertainty in interpretation due to equivalence is a fairly widespread feature of multilayer curves, its occurrence and importance depending in a complex way on bed thickness ratios and resistivity contrasts. It can only be resolved with the aid of additional geological or geophysical information.

An allied problem is suppression. For example, where three layers occur such that $\rho_1 < \rho_2 < \rho_3$ the middle layer may be so thin, or the resistivity contrast with one of the beds so small that its presence may not significantly alter the shape of the resistivity curve.

One assumption made in all calculations is that the strata are electrically isotropic. In fact many deposits conduct electricity better along the bedding planes than in the perpendicular direction, either due to preferential grain orientation or to the presence of thin alternating layers of low and high resistivity. A consequence of this anisotropy of resistivity is that depths calculated from resistivity curves may be significantly greater than true depths. However in many cases, particularly where the rocks are unmetamorphosed sediments, anisotropy is small enough not to have any serious effect.

So far the discussion has been confined to horizontal layers. Procedures for interpreting dipping contacts have been developed but in general depths to dipping strata can be satisfactorily interpreted using horizontal layer theory, provided the dips are moderate, say less than 20°.

So many factors control the accuracy of a depth interpretation that it is not possible to quote a figure for the likely accuracy. Very clearly it will be better in situations where there are a few electrically well contrasted layers and the approximation to the theoretical model is good. Control by a borehole at one point of the survey will help greatly, especially by resolving ambiguities due to equivalence. Where equivalence is not a problem and the data are good, an accuracy of 10 per cent may be achieved by careful and thorough measurement.

CHAPTER 5

Practice and Applications in Resistivity Surveying

5.1 Field Practice

WHATEVER the geophysical method being used to solve a particular problem the aim will be to acquire information quickly and cheaply consistent with obtaining enough data of sufficient accuracy for an adequate interpretation to be made. The choice of method will depend to a considerable extent on the geology. Resistivity methods are likely to be employed where the structures are simple and resistivity contrasts well marked, quantitative interpretation often being practicable only in situations in which no more than a few horizontal layers or one or two vertical interfaces are involved. Though investigation to depths of 1 km or more is possible such measurements require the use of very long arrays and interpretation tends to be particularly affected by lateral inhomogeneity. Both manmade surface obstacles and the complexity of the geology can therefore put severe constraints on practicable spread lengths, in some areas limiting the depth of investigation to under a hundred metres.

The type of electrode array used, the electrode spacing, the station pattern, the choice between depth sounding or constant separation traversing: decisions on all these matters have to be made at the planning stage, though they may well be modified during the survey in the light of experience. The more geological information there is about the area the easier it is to make the right decisions at the outset and it may well be economic to drill one or more exploratory holes to provide control even before the survey begins.

Where depths are to be determined in layered strata a series of soundings will be made using an expanding electrode system at each station. Only a single line of such probes with their expansion directions parallel to the strike may be necessary to determine the dip of a plane interface provided its direction is known. To survey irregular buried bedrock topography beneath unconsolidated overburden would require a grid of depth soundings. In carrying out such a survey we bear in mind the fact that quantitative interpretation will make use of plane layer theory. This will be inapplicable and results will therefore be almost meaningless unless the horizontal scale of the subsurface topography is several times that of the depth. A minimum distance between stations about equal to the bedrock depth is appropriate in such circumstances. Where the overburden is layered and laterally inhomogeneous borehole control is likely to be necessary.

Simple structures involving vertical or steeply inclined interfaces reaching nearly to the surface such as faults, igneous contacts, veins, brecciated zones, dykes or filled fissures and sinks in limestone may be investigated by means of constant separation traversing. Using appropriately small spacings the technique is also useful for mapping many archaeological features, e.g. walls, pavements and ditches. Apparent resistivity profiles calculated for simple structures are useful for comparison with measured profiles but quantitative interpretation is not normally warranted. In fact it may well be impossible due to lack of geological control and the complexity of the structure itself.

It is still possible to use constant separation traversing to detect structures such as faulted boundaries between different rock types beneath overburden providing the upper layers are not too variable in character and thickness. Otherwise their contribution to the total measured apparent resistivity will change laterally and the required "signal" due to the structure may be obscured. Constant separation traversing therefore has to be used with care and resort should be made to depth soundings if it is suspected that lateral inhomogeneity will obscure the interpretation of structure. As previously mentioned we may consider a constant separation traverse as sampling resistivity in a horizontal direction over a fixed depth range which depends on the spacing. Depth soundings provide information centred on one point about changes of resistivity with depth. If therefore we require to investigate a "two dimensional" structure, that is to say, one in which the cross section is constant in one horizontal direction (the strike), but which shows resistivity variation in both other coordinates, then to obtain the minimum information necessary for an interpretation a line of soundings perpendicular to the strike must be measured. Quantitative interpretation methods are only just being developed to deal with such problems but a useful picture of the distribution of resistivity in the ground across the structure can be deduced from a "pseudo-section". A line of soundings is measured perpendicular to the structure and the apparent resistivity values plotted beneath the station positions at depths proportional to the electrode spacing. The apparent resistivity is then contoured. Since depth of investigation increases with spacing the contour pattern crudely and qualitatively displays the resistivity variation in the ground. Figure 5.1 is a pseudo-section derived from dipole–dipole measurements made over a steeply dipping fissure zone carrying saline water in porous sediments otherwise saturated with fresh water. Habberjam and Jackson (1974) have developed semi-quantitative methods for interpreting such "resistivity spaces" when derived from Wenner soundings.

Both Schlumberger and Wenner configurations are commonly used for depth soundings, the choice being determined perhaps more by custom and preference than on scientific grounds. The Schlumberger array has a practical advantage in that it is not necessary to move the potential electrodes during a depth sounding until the potential difference becomes too small to measure accurately. Also, since local surface resistivity variations at the potential electrodes can cause irregularities in the apparent resistivity curve it is claimed that the Schlumberger array produces smooth data. Against this the Wenner array has the advantage that larger potential differences are being measured, the effects of local lateral inhomogeneities being to some extent averaged out. Both arrays have good resolving power in depth and about the same depths of investigation. The dipole–dipole array has a considerably greater depth of investigation than the Wenner array and is easier to use for very deep soundings since no cable connection is needed between the dipoles. Its disadvantages are that layer resolution in horizontally layered ground is not so good and measurements are more influenced by lateral resistivity changes.

Interpretation theory for apparent resistivity curves over layered strata assumes the layers are laterally homogeneous. If they are not, calculated depths and resistivities will be in error. It is important therefore to be able to test for lateral inhomogeneity when carrying out a sounding. One method of doing this is to employ the 5-electrode Lee Partition spread (see §4.5).

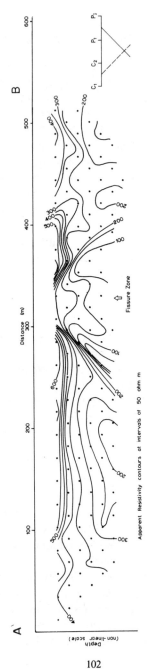

Fig. 5.1 A dipole–dipole "pseudo-section" over a steeply dipping conducting fissure zone in sandstones (reproduced by permission of J. W. Finch).

102

The fifth electrode is placed symmetrically between the two potential electrodes of a Wenner spread and at each station the left and right hand pair of potential electrodes is used for measurement. By this means two measurements of apparent resistivity are obtained from adjacent regions of the ground without moving the current electrodes, and therefore without altering the pattern of current flow within the ground between readings. These two values should both be the same, and if there is any difference between them it is due to lateral changes of resistivity within the length of the spread. A variation on this method has been described by Barker (1978) in which the five electrodes are equally spaced so that either of two Wenner arrays offset from each other by one electrode spacing can be selected by switching. Apparent resistivity measurements are then combined to average out lateral effects. Carpenter and Habberjam (1956) interchanged current and potential electrodes, giving three combinations, CPPC, CCPP and CPCP, and therefore three ground resistances which they designated R^{α}, R^{β} and R^{γ} (note incidentally that $R^{\alpha} = R^{\beta} + R^{\gamma}$ whatever the ground structure, thus providing a useful check on instrument accuracy). Using the correct geometric factor these resistances may be converted to apparent resistivities. If the ground is stratified and laterally homogeneous, though the three apparent resistivity curves generally cross each other (how often being determined by the number of layers and their relative resistivities) they will follow each other closely. However, if lateral changes occur in the ground the curves show sharp divergences and convergences. This tripotential method is most usefully used with the Wenner array where R^{α}, R^{β} and R^{γ} are of the same order of magnitude and thus measured with the same accuracy. Carpenter and Habberjam show how they can be combined to give a better approximation to the sounding curve that would be expected in the absence of lateral variations.

Sometimes it is necessary to investigate structures of small dimensions at depth, as for example cavities. It is important therefore to have some idea of the size to depth ratio at which such features become for practical purposes undetectable. In this context easily computed apparent resistivity curves for expansions over perfectly conducting buried spheres are illuminating. Figure 5.2 shows that the practical limit of detection is the point at which the depth to top is equal to the radius since in this case the resistivity drops at most by only 10 per cent from its normal value. This is an extreme case but as a rough guide we can apply the rule to a wide range of quite irregularly

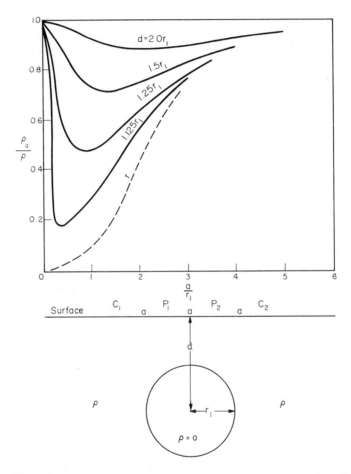

Fig. 5.2 Calculated apparent resistivity curves for an infinitely conducting sphere at different depths (modified from Van Nostrand and Cook, 1966).

shaped bodies provided the resistivity contrast is large and the host rock reasonably homogeneous.

5.2 Instruments

Resistivity instruments have to be portable, robust and weatherproof. Weight has therefore to be kept down, and so low power consumption is important. Current is usually supplied from dry cells or rechargeable batteries at between 20 and 100 mA depending on the depth range of the equipment, voltages at the current electrodes varying from 100 volts for shallow work to over 500 for deep investigations.

All methods of measurement make use of the four electrode principle already described. Current is fed into the ground between two electrodes, a potential difference being measured between the other two. This is expressed as the ratio of voltage drop to current and therefore has the dimensions of resistance. Since a very wide range of resistance has to be measured and potential differences can be very small (of the order of microvolts) high accuracy is required. If AC is used ground penetration depends on the frequency f used, the "skin depth" at which the signal amplitude is reduced by a factor of $1/e$ (i.e. to 37 per cent) being given by $Z_s = 500 \sqrt{\rho/f}$ metres for a resistivity of ρ ohm-m.

When shallow penetration of the order of 10 m is all that is required a frequency of 100 Hz is satisfactory. When the depth of investigation reaches 100 m frequencies of a few hertz are appropriate. For very large scale work it is necessary to use direct current, appropriate measures being taken to reduce the effect of natural earth currents which have very slowly varying DC components. Non-polarizing electrodes also have to be used with a DC source since electrochemical emf's are produced between the ground and metal electrodes. A simple electrode of this type consists of a small porous pot containing a copper electrode immersed in copper sulphate. Ground contact is made through the saturated walls of the pot and the solution. The use of alternating current avoids the problems created by the presence of spurious DC potentials for these alternately add to and subtract from the measured voltage and thus average out.

In many instruments, particularly those used for shallow investigations, interconnection between potential and current electrode circuits is required, so that all cables have to be brought to one point. When using large dipole–dipole configurations it is necessary to be able to measure current and potential separately to avoid the use of very long current cables. If this is to be done the current must be kept constant to a high degree of accuracy and some method employed for switching the potential measuring circuit in synchronism with the current circuit when this is reversed.

A method of measurement used in the ABEM Terrameter, is to compare the voltage drop across the potential electrodes with that across a potentiometer. The circuit is illustrated in Fig. 5.3. The output from the oscillator E is coupled to the current electrodes through a transformer T. In series with the current electrodes in the secondary circuit of T is a resistor R from which a signal V_r can be tapped by a sliding contact. In making a measurement, the amplifier A is first connected to the potential electrodes, and its gain adjusted so that the signal V_x gives a convenient reading on the meter M. The input of the amplifier is then switched to the signal V_r derived from the potentiometer R, and the sliding contact is

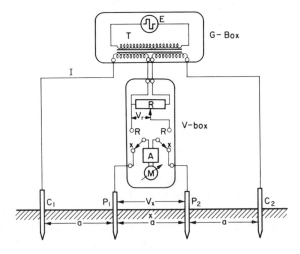

Fig. 5.3   Schematic circuit for the Terrameter (by permission of ABEM, Stockholm).

adjusted (using the same amplifier gain) until the same meter reading is obtained, so that $V_r = V_x$. Since the current through R and that through the ground are the same, it follows that the ground "resistance" V_x/I is equal to the resistance of that part of R from which V_r is derived.

Improved interpretation methods and the increasing use of computers have greatly speeded up data handling. This trend is likely to continue, making it important to increase the rate at which field measurements can be made with both improved range or accuracy. The introduction of digital circuitry has made this possible, the measurement cycle being completely automatic in a number of modern instruments. The digital version of the ABEM Terrameter provides a measuring range of 0.001 Ω to 1 MΩ and selectable periodicities of 1, 4, 16 and 64 s, the instrument averaging over 1, 4 or 16 cycles as required, resistances appearing on a 4 digit display. Automatic backing off of spontaneous ground potentials is included. An optional battery powered booster unit makes it possible to increase the output from 3 W to 50 W and thus greatly increase depth penetration.

For very deep penetration, particularly in low resistivity conditions Scintrex have developed a system using a separate receiver and transmitter. The latter is battery powered for shallow and intermediate depth work, motor generators of up to 15 kw being used for deep penetration. Current pulses of up to 8 s are available, current polarity cycling being automatic. The receiver is triggered by the transmitted pulse and locks on to the signal so that no cable connection between receiver and transmitter is necessary. Voltages down to a few microvolts are measureable. Transmitted current is held constant and monitored so that resistance can be calculated. The receiver incorporates circuitry for automatically backing off interfering natural ground currents. Advanced equipment of this type is also adapted for induced polarization measurements (see Chapter 7).

5.3 Applications

The resistivity method finds many applications in the fields of civil engineering and hydrogeology. In civil engineering more often than not the method is used to find the depth to bedrock beneath unconsolidated overburden and possibly the thicknesses and nature of the overlying sediments themselves. Bedrock depths must be known, for example, in sediment filled valleys where bridges, dams or motorways are to be built. The lithology of the overburden, whether clay, sand or gravel, may also be of considerable importance in site investigations. Mapping the extent and thickness of sands and gravels may also be carried out to assess reserves since such materials are of economic value. Since unconsolidated sediments can show a very wide range of resistivity, depending on degree of saturation, groundwater composition and porosity, borehole control for such surveys is very necessary.

Resistivity surveys have also been found useful for determining the extent and thickness of old spoil heaps.

In hydrogeology a very wide range of problems is met, demanding for their solution anything from a few depth soundings to a long and detailed survey. Major investigations are likely to involve also drilling, borehole logging and the use of other geophysical methods. In some instances the resistivity survey may be used only to determine the geometry of a known aquifer, that is, its depth and thickness, lateral extent and the degree and nature of any faulting. Quite often, however, and more particularly in unconsolidated formations, the survey may be a search for water bearing strata, in which case the properties of the sediments as well as their geometry are important. Clean sands and gravels, which have high porosities, made good aquifers. When saturated with fresh water they have medium resistivities (100–500 ohm-m) and can easily be differentiated from lower resistivity impermeable clays and marls, and from bedrock which is usually of much higher resistivity.

A characteristic situation, particularly in arid and semi-arid regions, is one in which porous sands overlie impervious bedrock, water being ponded in shallow depressions or in buried water courses. In theory this should give rise to a classical three layer curve with a low resistivity middle layer but the transition zone from saturated to dry conditions often introduces one or more extra layers and makes accurate determination of water table depth impossible. When interpreting such surveys it is important to have well logs available, particularly if the succession is complicated by the presence of, for example, silty or clay layers, since these may introduce ambiguities in interpretation, which have already been discussed under the heading of suppression and equivalence (see §4.7), effects which could not otherwise be detected and which lead to incorrect interpretation. An excellent example of this is given by Flathe (1970), and is shown in Fig. 5.4. The sounding was measured in the Lower Rhone Valley in the course of a hydrogeological survey, the objective being to map the depth of the very low resistivity impermeable layer. The sounding was measured close to a borehole, the succession being given in the figure. In the absence of additional information the curve would be interpreted as due to three layers, giving incorrect depths and resistivities. Though it appears to be a perfect three-layer curve it can equally well be interpreted using either four or five layers, the borehole data indicating that the five layer interpretation is in fact correct. Having established a proper correlation between borehole and resistivity data the interpretation of adjacent soundings could then be carried out correctly on the assumption that layer resistivities did not vary laterally.

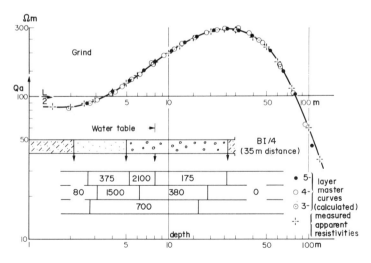

Fig. 5.4 An example of equivalence. Solutions for 3, 4 and 5 layers fit the sounding curve equally well within the limits of error. The log of a nearby borehole made a correct choice possible (after Flathe, 1970).

Groundwater is not always fresh. It may have picked up salt in travelling through a formation, or the salinity may be due to marine incursion. In the Netherlands the fresh water in the Pleistocene sands in many areas floats on a saline layer beneath. The conditions in one area are illustrated in Fig. 5.5 after Van Dam and Meulenkamp (1967). The water levels in the polder areas are artificially maintained at low levels, in fact well below sea level. Due to ditch cutting and other human interference the near surface peat and clay layers have been rendered semipermeable, resulting in an increased rate of leakage. In consequence there has been in many places an upconing of the lower saline layer, this even reaching the surface in places. Resistivity surveys here proved to be very useful for mapping the upper surface of the saline layer. It has also proved possible to use resistivities for calculating the salinity of the groundwaters, from which maps showing the chloride content in various regions have been prepared.

Use is now also being made of deep resistivity soundings in exploration for sources of geothermal energy. Since depths of investigation of a few kilometres are involved it is usual to employ dipole–dipole techniques,

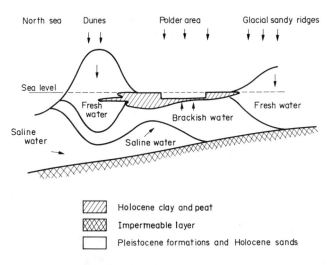

Fig. 5.5   The hydrogeological situation in the polder areas of the Netherlands (after Van Dam and Meulenkamp, 1967).

though other types of array have been used. In regions of geologically recent volcanic activity there are areas where the rocks at relatively shallow depth are still hot, the result of heat conduction from intrusive bodies beneath. These hot rocks are often heavily fissured and permit the circulation of groundwater. This reaches a high temperature and where the supply is adequate it can be tapped by suitably positioned boreholes. In fact temperatures may well be above 100°C, in which case the water flashes to steam as it rises to the surface. Natural hot springs are not uncommon in volcanic areas but sometimes impervious cap rocks prevent outflow and the geology may be such that even near the surface temperature measurements do not give unambiguous indications of thermal anomalies. However, fissured zones at depth, being saturated with water which is both hot and mineralized, have a low resistivity and can sometimes be mapped from the surface by electrical methods. Hatherton *et al.* (1966) describe a successful application of the method in New Zealand. Wenner constant separation surveys were made at 600 in and above and the data contoured. Using information available from other sources on average porosity at depth and groundwater conductivity it proved possible to calculate the likely variation of formation resistivity with temperature. Calculations indicated that apparent resistivities of less than 50 ohm-m indicated high temperatures at a few hundred metres depth, and in fact boreholes drilled on the basis of the geophysical indications all proved to be hot though not all were good steam holes.

CHAPTER 6

Other Electrical Methods

6.1 Electromagnetic Prospecting

6.11 *Principles*

ELECTROMAGNETIC methods of prospecting are based on the measurement of magnetic fields associated with alternating currents induced in subsurface conductors by primary magnetic fields. In most methods the primary or inducing field is artificially produced by passing an alternating current through a coil or loop but natural sources (e.g. the energy of thunderstorms) can be utilized. A major advantage of electromagnetic methods is the fact that many systems can be used from aircraft because conducting ground connections are not needed. Their main application is in the search for metallic ores, particularly sulphides. These are relatively good conductors, in marked contrast to their host rocks, and can thus be detected despite their localized nature and irregular shape.

The basic principle is most easily illustrated by reference to the two-coil system shown in Fig. 6.1. An AC current, usually with a frequency of between a few hundred and a few thousand hertz, is passed through the transmitting coil. Eddy currents are thus induced in the conducting ore body (and to a minor extent often in the host rock). In the diagram the ore body is represented as a vertical sheet and the induced current will have a more or less ring-like form in the plane of the sheet as shown. The magnetic field at the surface is the resultant of the secondary field created by this induced current and the primary field due to the transmitter. Usually there will be both a difference in direction and a difference in phase between the primary and secondary fields. The current induced in the horizontal receiving coil depends on the magnitude of the *vertical component* of the resultant field (and of course its frequency) and its

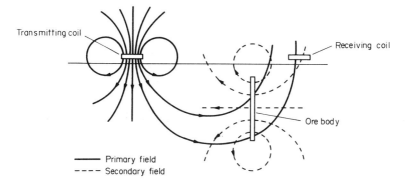

Fig. 6.1   Principle of the two coil electromagnetic system.

magnitude and phase may then be measured by comparison with the primary field. In practice an indirect way of doing this is adopted. Any sine wave signal can be separated into two components of the same frequency differing in phase by 90°, the individual components in general having different amplitudes. It is convenient to divide the signal from the receiver in this way into a component which is in phase with the signal generated in the receiving coil by the transmitter in the absence of a secondary field and an out of phase or quadrature part differing from this by 90° (see Fig. 6.2). These are also known as the real and imaginary parts of the signal.

Many factors control the phase difference between the primary and secondary fields, and therefore the relative magnitudes of the in phase and quadrature parts of the signal. This subject will be referred to later when interpretation is discussed.

To carry out a survey the two coils are separated by a fixed distance and traversed in line across the survey area. To ensure that there is no variation in the primary signal at the receiver the separation and coil orientation must not be altered between stations and where possible the coils must be kept at the same height. Where the ground is sloping and this is not possible a correction has to be applied.

The variations of the in-phase and quadrature components along a traverse over a vertical vein are shown in Fig. 6.3. In the absence of an

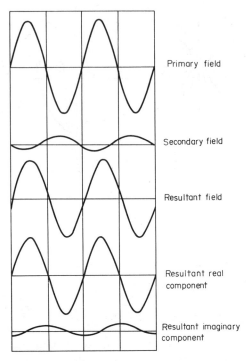

Fig. 6.2 Amplitude and phase relations for the two coil system. (a) AC primary field at the receiver; (b) secondary field at the receiver due to an ore body; (c) resultant of (a) and (b) picked up by the receiver; (d) real (in phase) component of the resultant; (e) quadrature (90° out of phase) component of resultant. The sum of (d) and (e) is equal to (c).

ore body the receiver picks up a constant in-phase signal due entirely to the transmitter. This primary field is given a value of 100 per cent and all increases or decreases resulting from the addition of a secondary in-phase component are expressed in terms of this. Since there is no quadrature field in the absence of a conductor the baseline for this component is zero, its amplitude also being measured as a percentage of the in-phase primary field. The form of the field curves can be understood by reference to Fig. 6.4. When the transmitter and receiver are on the same side of the vein the secondary field at the receiver reinforces that due to

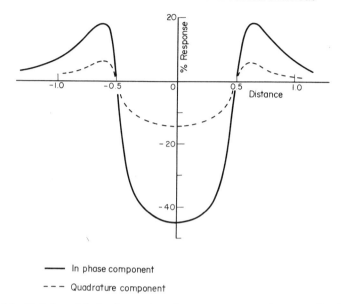

—— In phase component

– – – Quadrature component

Fig. 6.3   Variation of the in phase and quadrature components when traversing the two coil system over a vertical vein.

the transmitter. If the transmitter is directly over the vertical vein no flux is linked with the sheet and there is no primary field. Once the transmitter and receiver are on opposite sides the direction of the primary inducing field with respect to the sheet is reversed, with consequent reversal of the secondary field at the receiver. Primary and secondary field are thus opposed, reducing the real component below 100 per cent. For a vertical vein the flux lines on the surface over the vein are horizontal, so a second null occurs when the receiver crosses the body since it responds only to the vertical component of the field. Dip of the sheet leads to an asymmetry of the profiles, whilst variations in the geometry of the subsurface body inevitably change their general form.

When the geology is not too complicated some idea of ore geometry and attitude can be obtained by comparison of the field profiles with the results of traverses over models in the laboratory, or with data obtained by calculations. If the geometry is simple, as for example that of a

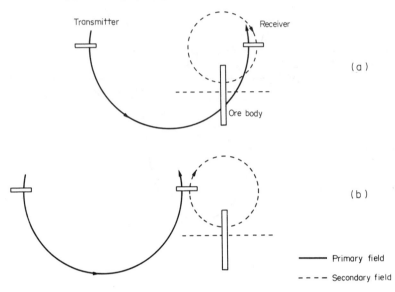

Fig. 6.4   The relative geometry of the primary and secondary fields when traversing a
vertical vein. Transmitter and receiver (a) on opposite sides, (b) on the same side.

dipping vein, the depth to its upper edge and the product of the
conductivity and thickness can also be deduced from the peak values of
the real and imaginary components measured on a traverse across the
vein. This is done by reference to a "phasor" diagram plotted from
laboratory measurements or from calculations for the appropriate model.
An example is given in Fig. 6.5. The imaginary component is scaled on
the vertical axis, the real on the horizontal. Two sets of lines are shown on
the diagram. One set gives values of the real and imaginary components
at the negative peak of the profile for equal depths to the top of the body,
but for a range of values of what is known as the "response parameter".
The radial set are lines of equal response parameter. For a semi-infinite
sheet the response parameter is the product of frequency, coil spacing,
conductivity, thickness and the magnetic permeability of the sheet (see
§8.1), but even for magnetic rocks this latter property has a value
little different from that for free space. Since frequency and coil spacing

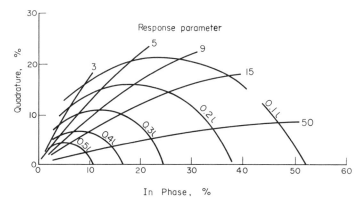

Fig. 6.5 Phasor diagram for the interpretation of the EM anomaly of a vertical semi-infinite conducting sheet.

are known, the conductivity-thickness product can be found, but not either quantity separately. This product has proved to be a useful measure of the economic value of this shape of ore body and it sets a lower limit to bodies worth investigating.

6.12 *Equipment and field technique—ground surveys*

In Fig. 6.6 the arrangement of the equipment used in the two coil moving source or Slingram system is shown. It is designed to be light and easily portable. A battery operated oscillator, usually providing two frequencies, of for example 800 and 3200 Hz, delivers current to the

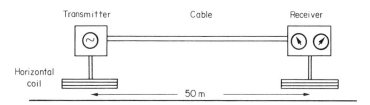

Fig. 6.6 Arrangement of the horizontal two coil or Slingram equipment.

transmitter coil. This is about 1 m in diameter and carried round the body, slung from the shoulders, or it may be a long solenoid with a highly magnetic core. The receiver, identical in design, is separated from the transmitter by a distance of between 25 and 100 m, the larger distances giving greater depth penetration. A reference signal is provided by the oscillator, against which the in-phase and quadrature parts of the received and amplified signal are measured, the magnitude being displayed on two meters as percentages of the primary in-phase signal at the receiver. A compensation method is used, the reference signal being mixed in opposition to the received signal until no sound is heard in a telephone headset in the circuit. The moving source system is light and portable and therefore easy to use, but its power is limited and so therefore is its depth of penetration. Practical depth of investigation depends on many factors such as size and conductivity of the ore body and on the "noise" caused by poor coil levelling or the effects of terrain, both of which may obscure the real signal. A figure of half the coil separation is usually taken to be the effective depth of penetration

The Turam fixed source method, the layout of which is shown in Fig. 6.7, has the advantage of greater depth penetration but the equipment is much less simple to operate. An insulated cable laid out in the form of a single turn square coil with sides up to 3 km in length is used as the transmitter. An AC generator supplies current at frequencies of 1–2 kHz depending on ground conductivity. Where the host rock, or more particularly the overburden, is significantly conducting lower frequencies have to be employed to obtain the necessary penetration. Parallel traverses are surveyed outside the loop, perpendicular to the cable. In this method not one but two horizontal coils, 10–50 m apart, in line and

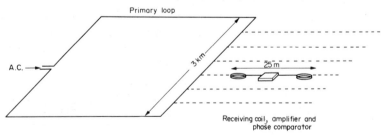

Fig. 6.7   Schematic layout for the Turam system.

moved together, are used as the detector. At each station the ratio of the amplitudes of the signals induced in the two coils and their phase difference are measured. For non-conducting ground the amplitudes can be quite simply calculated from the cable position and geometry, and phase differences are virtually zero. It is usual to normalize observed amplitudes by dividing their ratio by the calculated ratio for a non-conducting earth so that in the absence of an ore body normalized amplitude ratios are unity.

Interpretation proceeds in a somewhat similar manner to that used for the moving source method, by comparison of observed and calculated or model profiles. Amplitude ratio and phase differences provide the same level of quantitative information.

Electromagnetic methods other than the two described have been developed in which different coil arrangements are used. In one relatively simple but effective technique only the azimuth and tilt of the resultant field direction (strictly speaking, that of the major axis of the "polarization ellipse") are measured, the tilt being an indicator of the presence of a subsurface conductor. In this "tilt angle" method the transmitting coil is commonly used in the vertical position, with its axis parallel to the traverse line. The receiving coil is set up in the plane of the transmitting coil and rotated about the horizontal to find the direction of minimum signal, this giving the tilt of the resultant field direction. In one version of the field technique transmitter and receiver are moved together along parallel traverses, the tilt at each station being measured. Figure 6.8 shows the arrangement.

6.13 *Airborne electromagnetic methods*

Electromagnetic methods share the great advantage of magnetic and radiometric methods that they can be adapted for use in either fixed or rotary wing aircraft, though here again the ability to gather data rapidly to a set pattern, whatever the difficulties of the terrain, has to be set off against a height penalty which results in reduced resolution and ground penetration. Depending on conductor size and shape the receiver signal strength falls off at a rate of between the fourth and the sixth power of the height. Good accounts of these methods are given in articles by Pemberton (1962) and Ward (1967).

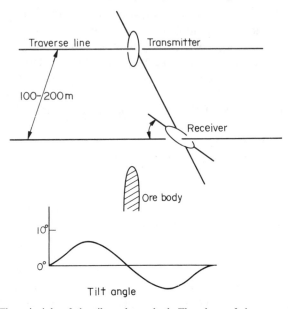

Fig. 6.8   The principle of the tilt-angle method. The plane of the transmitter coil is vertical and at right angles to the line of traverse. The receiver coil is rotated about a horizontal axis perpendicular to the line. Both coils are moved together.

In methods in which a moving source is used source and detector can be mounted on the same aircraft (e.g. on the two wing tips). If a helicopter is used transmitter and receiver coils are attached to opposite ends of a 10–20 m horizontal boom suspended on a cable beneath the machine as in Fig. 6.9(a). Relative coil orientation depends on the system used, coils being vertical or horizontal, coaxial or coplanar. An advantage of rigidly fixed coil systems is that with careful mounting spurious signals due to relative coil motion brought about by wing flexure and vibration can be kept down to a few parts per million of the field at the receiver. Equally important, both real and imaginary components can be continuously recorded. A limitation is the small coil separation, this leading to a relatively low signal amplitude for steeply dipping conductors and a reduced sensitivity due to a decrease in the signal/noise ratio. Theoretically   response   can   be   improved   by   increasing   the

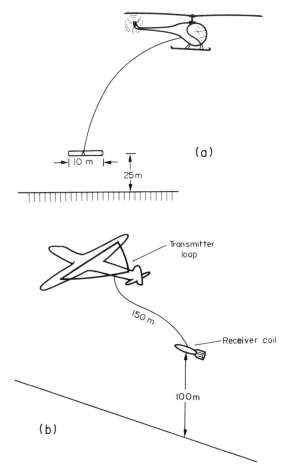

Fig. 6.9   Airborne EM systems. (a) The transmitter and receiver are mounted at opposite ends of a boom slung beneath the helicopter. (b) The transmitting coil is fixed to the aircraft and the receiver towed behind it from a cable.

transmitter/receiver separation, one way of doing this being to tow the receiving coil in a streamlined container at the end of a 150 m cable as in Fig. 6.9(b). As it then becomes impossible to maintain the relative orientation of the two coils with high accuracy fluctuations in the large in-phase primary signal in the receiving coil occur. These are large enough to obscure the relatively small real part of the secondary signal due to the ore body and so only the imaginary part can be measured. The imaginary part of the signal is not affected by the transmitter since, as has been explained, the primary signal has no imaginary component. To make up for loss of information, measurements are made at two frequencies, in one system these being 400 Hz and 2300 Hz. There has been considerable controversy about the relative merits of fixed and towed systems but the former now seem to be much more used. The rapid fall off of anomaly with height necessitates as small a flying height as possible compatible with safety, this being about 150 m for fixed wing aircraft, giving perhaps 100 m effective ground penetration. The small size of many ore bodies also makes it important to fly profiles spaced only a few hundred metres apart or even less when more detailed work is being done with helicopters. Accurate position and height control is of course imperative.

Interpretation proceeds very much along the lines used for ground electromagnetic methods but in many instances may not be carried beyond the stage of qualitative comparison of field and model or calculated curves. The information obtained in, at the most, a few passes at considerable height above the ore body is much more limited than that gained from a close network of profiles measured at ground level. Many anomalies due to fault zones and other conductive bodies of no economic interest are often observed, and the first stage of interpretation is to eliminate as many of these as possible.

A major problem of airborne electromagnetic methods is the detection of a signal of strength perhaps no more than a hundred parts per million of the primary field at the receiver in the presence of "noise". One important source of noise in this context we have already noted, i.e. rapid random variations in the primary field of the same order of magnitude as the true signal, as a result of relative transmitter receiver motion due to vibration and wing flexure. There are also other sources, such as those due to currents induced in the metal body of the aircraft. The INPUT system, developed initially by Barringer, avoids many of the noise problems by using a pulsed source and measuring the decay of the signal induced in the ground conductor after the energizing current has been switched off. A large horizontal three-turn coil is used for the transmitter, the receiver being a much smaller vertical coil towed 150 m behind the aircraft. The energizing signal is of half sinewave form, pulses being alternately positive and negative and having a duration of one or two milliseconds (Fig. 6.10). The ground signal is sampled six times in the off period between pulses, the six samples being fed to separate channels and individually recorded. In addition to the information obtained from an analysis of the variations along the traverse of the peak amplitude of the signals from a single channel, signal duration and fall off, noted by comparing relative channel amplitudes for each transmitted pulse, is also

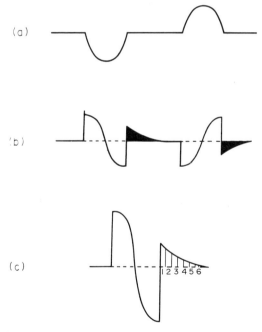

Fig. 6.10 The INPUT system. (a) The form of the primary magnetic field. (b) The signal generated in the receiving coil. Note the transient due to the ore body which continues after the cessation of the primary pulse. (c) The signal in the receiving coil is proportional to the rate of change of the field.

diagnostic. As well as proving a satisfactory system for the detection of highly conducting ore bodies the method appears to be useful in the detection of flat lying conductors such as aquifers, especially where there is a need to differentiate between fresh water areas and those affected by saline incursions.

6.14 *Passive electromagnetic methods—AFMAG and VLF*

Two other systems need mention, which can be used either on the ground or in the air. They differ from those so far described in that a distant power source not under the control of the operator is used as the transmitter. In the AFMAG (Audio Frequency Magnetic Field) system this is the energy of distant thunderstorms, which generate an electromagnetic field as shown in Fig. 6.11, the electric component being parallel to the generating lightning flash.

The space between the earth surface and the ionosphere acts as a wave guide and at great distances, in the absence of a conductor, the magnetic

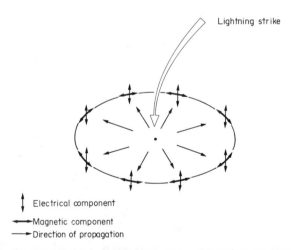

Fig. 6.11   The form of the electromagnetic field at the earth's surface at a great distance from a lightning strike. The electric component of the field is vertical and the magnetic component horizontal.

part of the field is therefore horizontal. Amplitudes vary randomly and the energy arrives from all directions, though the azimuths do sometimes show some degree of order when measured over a period of time. In the presence of a conductor a secondary field is generated, the resultant being tilted and showing a tendency to concentrate in one azimuth. In the airborne system tilt angles are calculated from the combined signals received by two identical orthogonal coils towed in a housing at the end of a 70 m cable. The axes of both coils are set up at 45° to the horizontal so that no signal is received when the field is horizontal. Recording is continuous, the signal being integrated over a period of a few seconds for smoothing purposes. As in the ground based tilt angle method of prospecting already described the information is plotted to show the dip of the field along the line of traverse. The method has proved more useful for the detection of geological structures involving relatively small conductivity changes than for finding localized highly conducting ore bodies. The frequencies of a few hundred hertz that are used give good depth penetration.

VLF (Very Low Frequency) systems are in many respects similar to AFMAG but make use of the energy of distant and powerful radio transmitters in the 15–25 kHz range rather than natural radiation. The field here again is one in which the electric component is vertical and the magnetic horizontal (see Fig. 6.11). Various ways of detecting the presence of secondary fields are adopted. Since such fields, unlike the primary, have a vertical in-phase and a quadrature component both these can be measured, the vertical electric field, which is less affected by subsurface conductors, being used as reference. This is the basis of airborne methods. Ground equipment is very light and portable. In one instrument the receiver consists of two orthogonally mounted coils. After orientation in the correct azimuth the equipment is rotated about a horizontal axis, the point of minimum signal from the "vertical receiving coil" giving the tilt of the resultant field. The primary signal in the other coil is used to provide a quadrature reference signal for backing off (and thus measuring) the vertical component of the quadrature signal in the receiving coil.

VLF transmitting stations provide practically world-wide coverage for the system with greater reliability than the storm sources of AFMAG. The higher frequency means that the penetration is shallower, but the system, like AFMAG, has been found useful for mapping concealed boundaries between formations of contrasting resistivities rather than for the detection of localized conductors.

6.2 Induced Polarization

6.21 *Principles of the method*

The average electrical conductivity of a disseminated ore body, in which only a small percentage of the ore particles are in contact, is likely to be much lower than that of a more massive body of similar size. Consequently disseminated metallic ores are difficult to detect by electromagnetic or resistivity methods. However, in the presence of water, contained in pores and fissures, when a direct current is passed certain classes of ores in a disseminated state show the effect now known as Induced Polarization (IP). Electrochemical reactions take place and electrical energy is stored. After switching off the current the stored energy is discharged, causing a current to flow through the surrounding ground which is detectable at the surface.

The phenomenon is shown schematically in Fig. 6.12(a). In the figure, for simplicity of illustration, a Wenner array is shown but in practice the dipole–dipole arrangement is normally used. The ore is shown to be electrically charged as a result of the passage of current through the electrodes C_1 and C_2. The dotted lines illustrate the discharge that takes place once the primary current is switched off.

Figure 6.12(b) shows the resulting changes in voltage during one cycle of measurement. On switching on there is an instantaneous rise of the voltage to a particular value. After this follows a slow increase to a steady state. On switch off the potential falls instantaneously by an amount equal to the original rise, after which follows a decay to zero, this taking seconds or even minutes.

The form of the discharge curve V_t can be described by the sum of several exponentials $V_t = V_o \exp(-t/r)$ with different time constants r, two being adequate to cover the period between 1 s and 10 s. In this "time domain" type of IP measurement, such a measurement cycle is followed by one of reversed polarity.

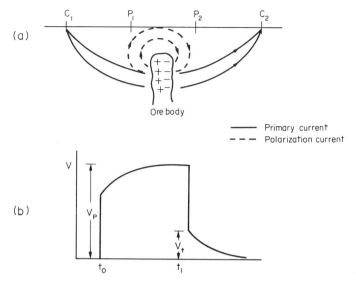

Fig. 6.12   The Induced Polarization Effect. (a) Flow lines of primary and polarization current in the ground for a polarizable ore body. (b) Measured voltage V_p between P_1P_2 for time t_0 to t_1, V_t being the IP voltage.

The induced polarization therefore opposes both the build up and the collapse of the primary potential difference and is sometimes referred to as the overvoltage, for an additional potential over and above that required to overcome the ohmic resistance is required to drive current through the ore. Note that this secondary voltage V_t which has to be overcome on switching on the current is the residual value to which the potential falls at switch off. Now the IP phenomenon behaves linearly so that any change in current and therefore primary voltage causes a corresponding change in IP. Consequently V_t/V_p the ratio of the instantaneous value of the residual potential difference to the primary voltage is independent of ground current and is theoretically a good measure of IP. In practice electromagnetic induction in the ground takes a short time to die away and V_t is not measured until some tens of milliseconds after switching off the energizing current. Nor in fact is the primary voltage normally given time to reach its steady state, but this does not matter provided times are kept constant for all measurements. The ratio V_t/V_p is known as the chargeability and of course depends on the time at which V_t is sampled. A facility is also provided for sampling the transient several more times during the decay period.

In a qualitative sense the ore behaves as a very large capacitor in which electrical energy is stored. Thus if AC rather than DC is passed into the ground the ratio of voltage to current measured becomes an "impedance" rather than a resistance and a change in value with frequency is to be expected. This is in fact so, but most of this change takes place between DC and 10 Hz, far too low in the frequency range for the effect to be due to the dielectric properties of the rock but not unreasonable if the energy storage mechanism is electrochemical. The IP effect can therefore also be measured using alternating current; this is the "frequency domain" approach. In theory the ground impedance is measured at zero frequency (i.e. DC) and at a specified AC frequency. Apparent "resistivities" are calculated in the normal way from the measured values of V and I and the electrode separations. A quantity known as the *percentage frequency effect* is then defined, this being

$$FE = 100\,(\rho_{DC} - \rho_{AC})/\rho_{AC}$$

In practice, to simplify instrumentation, a measurement at very low frequency (e.g. 0.1 Hz) is substituted for the DC measurement, the second being made a decade higher (1.0 Hz). We shall continue for simplicity to write of the DC measurement when in fact the very low frequency measurement is intended. Strictly, therefore, the IP measure just described should be called the Partial Frequency Effect since it is measured over a limited range of the frequency spectrum.

Remembering that conductivity is the reciprocal of resistivity the Frequency Effect may be written as

$$FE = 100(\sigma_{AC} - \sigma_{DC})/\sigma_{DC}$$

Another measure of induced polarization, derived from the Frequency Effect, is the Metal Conduction Factor. It is defined as

$$MF = FE \times \sigma_{DC} \times 2\pi.\,10^5$$

The constant multiplying factor has no fundamental significance. Apart from this constant the MCF is equal to $(\sigma_{AC} - \sigma_{DC})$ and is sometimes considered to be a better measure of the amount of ore present than the Frequency Effect.

6.22 *Origin of induced potential*

The origin of the IP effect is not yet completely understood. A model of an ore body together with a much simplified equivalent electrical circuit is given in Fig. 6.13. As described in Chapter 4 conduction in the ground takes place through the electrolyte in the connected pores and fissures of the rock and is therefore ionic. Many of the conduction paths in a disseminated ore body are blocked by mineral particles in which the carriers of current are electrons. Complex reactions take place at the grain surfaces and on passing a current ionic equilibria are disturbed, leading to a build up of positive and negative charges at opposite surfaces and the subsequent polarization of the ore. Each grain, in fact, acts qualitatively as a capacitor, though the time constant of discharge is far too long for the effect to be a dielectric one. On switching off, the ions diffuse back to their equilibrium state causing a decaying current to flow which has a direction in the ore opposed to, and outside the ore in the same direction as the original energizing current so that the potential difference measured at the surface is maintained.

Since each blocked pore becomes polarized the total IP will be related

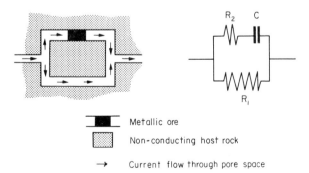

Metallic ore

Non-conducting host rock

→ Current flow through pore space

Fig. 6.13 Schematic model of a disseminated metallic ore body together with simplified equivalent electrical circuit to illustrate the IP effect. R_1 is resistance of unblocked pore, R_2 resistance and C capacitance of blocked pore.

to their number and therefore not only to the volume of the ore but to its state of dissemination. Very small grains, however, offer a very high surface resistance so that the energizing current in a finely disseminated ore will travel preferentially through unblocked pores, setting a limit to the IP amplitude. High porosity and high groundwater conductivity also reduce IP for both lead also to a short circuiting of energizing current through unblocked paths.

Though there is a difference between the degree of polarization of different ore species and to some extent between the form of the decay curves, many irrelevant factors control the measured IP amplitude and it has not proved possible to discriminate between different ores using the methods described above. More recent techniques in which the response in both amplitude and phase is measured, not at two discrete frequencies but over a whole range, may prove more successful in this respect. Such techniques known as "complex resistivity" have bridged the gap that has hitherto existed between the resistivity and IP approaches.

Clay bearing sediments also show weak IP. The effect is connected with the movement of adsorbed cations on the clay lattice. As yet only very limited use is made of the phenomenon but it may well find some applications in the groundwater field.

6.23 *Equipment*

Basically IP equipment is simple but automation and other improvements that have taken place over the last few years have made the apparatus quite complex. In light weight systems power is derived from batteries. Equipment designed for deeper penetration and thus requiring more power makes use of AC generators of up to a few kilowatts and rectifiers supplying a constant DC current. In time domain apparatus the DC current is automatically chopped and reversed at a fixed rate, as for example 2 s on and 2 s off before reversal, giving an 8 s cycle. Reversal and integration over a number of cycles improves signal to noise ratio and averages out the effect of other ground currents. In modern instruments of this type the receiver does not have to be connected to the transmitter. It has a programmed measurement cycle which is initiated by a sensing device triggered by the current in the ground. Sampling of the decay curve is done in slightly different ways in different instruments. In the Huntec instrument this is done at four geometrically spaced intervals, the changing voltage being sampled over a short time interval at each sampling point. The instrument continues to cycle, averaging chargeabilities, until stability is achieved. The four values at different points on the decay curve are displayed on digital voltmeters.

Various systems have been adopted for the measurement of frequency effect. It is now normal for the necessary measurements of the apparent resistivity at the two frequencies to be made simultaneously. The McPhar system does this by transmitting a dual frequency waveform, the two frequencies being separated in the receiver and their amplitudes measured. Frequency effect is then automatically calculated. Scintrex has developed equipment in which a low frequency square wave is transmitted. Since a square wave is a synthesis of all odd harmonics it is possible by using appropriate filters to separate out the fundamental frequency f and third harmonic $3f$ and measure the frequency effect between the two.

6.24 *Field procedure and interpretation*

The standard approach to IP surveying for ore bodies is to lay out a set of parallel traverses across a prospect perpendicular to the structural trend. The dipole–dipole array (see §4.5) has a number of advantages over other configurations for this kind of surveying. It is easy to use since no connection is required between dipole pairs, and in addition separation of current and potential cables cuts down electromagnetic coupling, thus reducing the amplitude of any spurious signal. Dipole-dipole arrays are also more sensitive than other electrode configurations to lateral changes in the electrical properties of the ground, such as are being sought in the search for ores.

A pseudosection (see §5.1) is then built up from the calculated FE's or chargeabilities. It is also usual to measure resistivity at the same time.

Figure 6.14 gives an example of data from a frequency domain IP survey over a sulphide ore body. The values plotted are of the Metal Factor.

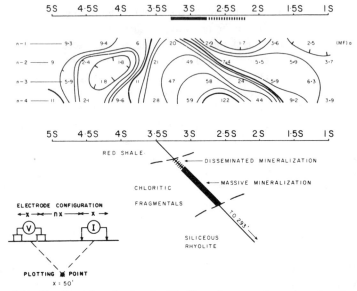

Fig. 6.14   Pseudo-section showing the IP effect of a zone of massive mineralization, Western Australia. The quantity plotted is the metal factor (courtesy of McPhar Geophysics Ltd.)

The position of the ore body is indicated by the high values of the contours but the vertical scale is quite arbitrary. Nor in fact is the pattern of the contours a true indication of the shape of the ore body, a vertical vein for example giving rise to a triangular area of high IP, the contours widening downward. This is most simply explained by saying that increased electrode separation gives rise not only to a greater depth but also a greater radius of investigation. The increased value of IP due to a body "seen", say, beyond one end of a spread is moreover plotted directly beneath its centre. "Electrode effects" of this kind can be complex.

Ore bodies can have complex shapes, and the host rocks in which they occur are often inhomogeneous. What is more, once a prospect has been detected it has to be drilled. There has not therefore in the past been much incentive to develop quantitative methods of interpretation of IP surveys but improvements in computational techniques and field methods have of late led to a good deal of interest in this subject.

CHAPTER 7

Gravity Surveying

7.1 Introduction

THE aim of gravitational prospecting is to detect underground structures by means of the disturbance they produce at the surface in the earth's gravitational field. Though the method is fundamentally a simple one the field differences to be measured are so small that in practice the instruments used and the techniques employed are highly sophisticated.

The basis of the method is Newton's Law of Gravitation. This states that every particle of matter exerts a force of attraction on every other particle, this being proportional to the product of the masses and inversely proportional to the square of the distance between them, i.e.

$$F = G \frac{m_1 m_2}{r^2} \qquad (7.1)$$

where F is the force between the two particles of mass m_1 and m_2, r is their separation and G the universal gravitational constant. Its accepted value, which has been obtained by direct experiment, is 6.67×10^{-11} $N m^2 kg^{-2}$.

By the second law of motion the acceleration a of a body of mass m_1 due to the attraction of a mass m_2 is given by

$$a = \frac{F}{m_1} = G \frac{m_2}{r^2} \qquad (7.2)$$

The force per unit mass acting on the body is therefore equivalent to its acceleration and is known as the gravitational field at m_1. If we put m_2 equal to the mass of the earth and r equal to the earth's radius the resulting acceleration is roughly equal to the gravitational acceleration on

133

the earth's surface and is directed radially downwards. However, the attraction due to, say, a small buried mass will act along a line joining the centre of the mass to the point of measurement. If the mass is *dm* then the acceleration

$$a = G\frac{dm}{r^2} \qquad (7.3)$$

The instruments used for gravitational surveying are known as gravity meters or gravimeters and are designed to measure field differences between localities, not absolute values. To make a measurement the axis of the instrument is aligned in the field direction, i.e. perpendicular to a plane defined by spirit levels on its upper face. The gravitational field direction is, of course, the direction of the resultant of the earth field and that due to local masses. Such local fields are, however, relatively very small indeed compared with the earth's gravity so that the direction of the latter is virtually unaffected, though its magnitude changes very slightly. The direction of the "vertical", the direction of the sensitive axis of the gravimeter, is then everywhere the same for all practical purposes. Therefore, in calculating the effect on a gravimeter of a local subsurface mass it is the vertical component of its field that has to be considered. At any point on the surface this component is

$$g_z = a\cos\theta = G\frac{dm\cos\theta}{r^2}$$

$$= G\frac{zdm}{r^3}$$

$$= G\frac{zdm}{(x^2 + z^2)^{3/2}} \qquad (7.4)$$

where z, r and θ are as defined in Fig. 7.1.

If the body is a large one, then its total effect is

$$g_z = G\int \frac{zdm}{r^3} \qquad (7.5)$$

where the integral sign indicates that the total field is obtained by summing the effect of all elements throughout the mass. When the body

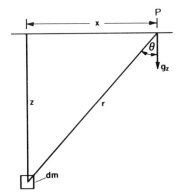

Fig. 7.1   The gravitational effect of a mass element.

has a simple geometrical shape an exact expression for the integral can be obtained. Otherwise the total is obtained numerically by dividing the body into small elements, calculating separately the attraction due to each, and then finding the sum.

In survey work modern practice is to measure anomalies in "gravity units" (μN kg^{-1} or μm s^{-2}). The older unit, the milligal, still appears in recent literature and is equal to 10 g.u.

The earth's gravity field is roughly 10^7 g.u., gravity anomalies due to geological structures of interest varying from a few gravity units to perhaps a few thousand. It is now possible to measure spatial changes in the field to better than $1:10^8$ but before interpretation in geological terms can be considered it is necessary to separate out the gravitational effects of subsurface density changes from those with a different origin. The methods adopted for doing this are described in the next section.

7.2 The Bouguer Anomaly

That part of the difference between observed gravity and theoretical gravity at any point on the earth which is due purely to lateral variations of density beneath the surface is known as the Bouguer anomaly. To obtain this quantity observations have to be corrected to allow for changes in gravity with latitude and height and for the attraction of topography. These corrections are discussed below.

7.21 *Latitude correction*

If the earth were a homogeneous (or concentrically layered) non-rotating sphere with the same vertical density gradient everywhere, apart from local near surface density variations due to geological structure, and if it had a surface parallel to sea level, then clearly all gravity variations over the surface would be caused by geological structure. But this is not so. The best approximation to the shape of the earth for practical purposes is an ellipsoid of revolution with an equatorial radius (6378 km) about 20 km greater than the polar radius. This is known as the reference ellipsoid and is virtually the sea-level surface. Because of the flattening the poles are nearer to the centre of mass than the equator, so that gravity increases with increasing latitude. This effect is increased by the opposing acceleration due to the earth's rotation, which has a maximum equatorial value of about 1/3 per cent of the gravitational attraction. The variation of gravity with latitude over the surface of an ellipsoidal earth can be expressed in the form

$$g = g_0(1 + C_1 \sin^2 \phi - C_2 \sin^2 2\phi) \qquad (7.6)$$

where g_o is the value of gravity on the equator and ϕ is latitude. C_1 and C_2 are constants which depend on the earth's shape, the numerical values of which have been adjusted to give a best fit to the measured variation of gravity over the earth's surface. With these values equation (7.6) becomes

$$g = 9.780318(1 + 0.0053024 \sin^2 \phi - 0.0000058 \sin^2 2\phi) \, \text{m s}^{-2}$$

This is the 1967 International Gravity Formula. It gives the value of gravity at any latitude on the reference surface. From it can be calculated the corrections that must be applied to the measured values of a survey to allow for the increase of gravity from the equator to the poles. Since in prospecting we are often concerned only with gravity differences the correction may be applied as a difference from an arbitrarily chosen base and, as the rate of change of gravity with latitude is almost linear over a small range, provided the survey is of limited extent a constant factor can be used. In mid-latitudes this works out to be roughly 1 g.u. per 100 m north or south.

7.22 *Elevation correction*

Since gravity varies with height it is necessary to correct all observations to a datum which is usually but not always sea level (i.e., the surface of the reference ellipsoid). This correction consists of two parts, the first of which is known as the free air correction.

Provided we assume the earth to be a sphere the mass can be considered concentrated at its centre and the value of gravity at sea level is given by

$$g_o = G \frac{M}{R^2} \tag{7.7}$$

where M = mass of the earth,
$\quad R$ = its radius.

The value of gravity g at a height h above this is given by

$$g = G \cdot \frac{M}{(R + h)^2} = G \frac{M}{R^2} \left(1 - \frac{2h}{R} \dots \right)$$

Therefore, ignoring higher order terms

$$g = g_o \left(1 - \frac{2h}{R} \right)$$

or

$$g_o - g = \frac{2g_o h}{R} = Fh \tag{7.8}$$

In calculating this expression the fact that the earth is ellipsoidal rather than spherical has been neglected, but the effect of this simplification is negligible. In fact in equation (7.8) we can replace g_o by g_m, the mean value of gravity over the earth's surface, and R by the mean radius R_o. The resulting constant factor F by which h is multiplied to obtain the gravity correction can now be used when working in any latitude and at all but exceptional altitudes and has the value 3.1 g.u. m^{-1}.

The factor is of course the gradient of gravity in free space. As gravity decreases with height this correction has to be added to the observations to correct to sea level. The free air gradient, however, accounts for only part of the change in gravity with height. Between any elevated station and sea level there is a thickness of rock exerting a gravitational attraction at the surface which can be considered to be additional to that due to the mass of the earth below the ellipsoid. To a first approximation its attraction can be taken to be that of an infinite slab of thickness equal to the station height. The attraction of such a slab is $2\pi G\rho h$ where h = station height and ρ = density. When reducing an observation down to sea level the correction has to be subtracted since we are removing from beneath the station a slab of rock of thickness h and consequently reducing the downward attraction by an amount $2\pi G\rho h = \beta\rho h$. This correction is known as the Bouguer correction after the French geodesist who first made use of it. It is usual to combine the free air and Bouguer corrections into a single elevation correction of the form

$$g_o = g_h + (F - \beta\rho)h \qquad (7.9)$$

where $\beta = 0.42$ g.u. t^{-1} m²
   g_o = gravity corrected to sea level
   g_h = observed gravity at height h m.

Assuming that the average rock density is 2.5 t m⁻³* the total correction amounts to roughly 2 g.u. m⁻¹ so that heights must be known to better than 0.5 m to retain a relative accuracy of 1 g.u. between stations.

   A further problem in making the Bouguer correction is to know what density to use. For example, an error of 0.05 t m⁻³ density for a station 500 m above sea level leads to an error of over 10 g.u. in the Bouguer correction. For a survey of any size a geological map is a necessity, for this gives the surface boundaries of the various rock formations. Densities are then found by laboratory determinations on a representative collection of rock samples and also, where possible, from a statistical analysis of field data. For example, in any area where the density of the rocks is approximately constant, a set of gravity observations may be taken over a small topographic feature. The observations are corrected

---

*See §1.2 for a note on the units of density.

for latitude effect and free air gradient and, assuming for the moment that there is no geological structure present causing gravity variation, the corrected values will be due only to the Bouguer effect of the topography. Thus if the corrected data are plotted against height the resulting graph will be a straight line, the gradient of which will be $\beta\rho$ g.u. m^{-1} so that the density can be determined. The range of rock densities encountered is from about 2.0 to 3.0 t m^{-3}, although exceptional materials such as peat (1.0 t m^{-3}) and ore bodies (3.0–4.0 t m^{-3}) lie outside the range. Some typical values are given in Table 7.1. In practice in any area there is almost

Table 7.1   The densities of some common rock types. Sandstone and limestone densities are to a large extent determined by porosities which may be as much as 30 per cent.

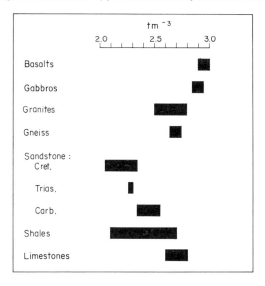

always some gravity variation due to geological structure but unless this happens to be correlated with topographic height it will produce merely a random scatter of the plotted points. Even so a reasonably accurate value of the density can usually be obtained from the straight line which best fits the data.

7.23 *Terrain correction*

When the topography is relatively flat the elevation correction may provide a sufficiently accurate method of reducing the data to sea level or any other convenient datum. If there are considerable irregularities of elevation, particularly in the vicinity of the station, then the simple assumption that there is an infinite slab of rock between the observation point and sea level is inadequate and a further allowance must be made for departures from this. Graphical or, more usually, computer methods are used to calculate the gravity effect of all hills above the station height, together with the mass deficiencies caused by valleys, these having been assumed to be rock filled in the Bouguer slab correction. Hills and valleys will both lead to a positive correction to the observed gravity. One approach (Bott, 1959) is to divide the topography into kilometre square columns and feed to the computer average surface elevations and positions for each column, together with gravity station heights and positions. Values of rock density are also assigned to the various rock columns. Using an appropriate formula the gravity effect of the rock columns round each station out to a pre-determined radius is calculated and the total terrain correction obtained. A hand method using specially designed transparent graticules is described by Hammer (1939). The principle is the same.

7.24 *Definition of the Bouguer anomaly*

The gravity values can now be presented as Bouguer anomalies, these being defined as the discrepancy between observed gravity and that expected on allowing for all known effects, i.e.

Bouguer anomaly = observed gravity + elevation correction + topographic correction − theoretical gravity on the reference ellipsoid at the same latitude.

Because only gravity differences will have been measured the survey must start from a point of known gravity if the true Bouguer anomaly is to be presented. In prospecting we are often only interested in local differences of gravity anomaly, and this being so it is quite satisfactory to start from any convenient base to which an arbitrary gravity value is given. To observed differences from this point are then applied the elevation and topographic corrections and a correction for the difference in gravity due to the latitude change from base. The resulting anomaly values differ from the true Bouguer anomaly only by a constant amount.

It may also be necessary if high accuracy is sought to apply what is known as a tidal correction to the observations. The attraction of the sun and the moon results in a cyclic change in gravity which may be as much as 3 g.u. in 6 hours. The calculation is complex and the correction can be obtained from pre-published tables for the year.

7.3 Instruments

All measurements of gravity for prospecting purposes are made with gravimeters. These are designed to measure directly small differences in the strength of gravity and are quickly operated portable instruments with a sensitivity quite adequate for all survey purposes. Modern instruments are based on the same principle as the long period vertical seismograph. An approximately horizontal beam hinged at one end carries a weight as shown in Fig. 7.2. The beam is connected to the mainspring which is attached at its upper end to a support directly above the hinge. The moment the spring exerts on the beam is Sa where S is the restoring force in the spring and a is its perpendicular distance from the hinge. This balances the gravitational moment $mgl \cos \theta$, θ being the

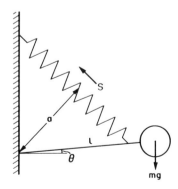

Fig. 7.2   The principle of the astatic gravimeter.

small angle the beam makes with the horizontal. If gravity increases the beam will deflect downwards until a new equilibrium position is reached. This will cause an increase in the spring length and hence of the restoring force *S* but the distance *a* will decrease. It is possible, therefore, by proper selection of the point of attachment to arrange for as small an increase of the restoring moment with increasing gravity as is desired. Using an ordinary coiled spring such an arrangement would have a very small stable working range. However, by taking advantage of the properties of pretensioned ("zero length") springs it is possible to produce an instrument with a linear and very sensitive response over a wide range.

Such gravimeters are described as astatic. They behave as if there were two separate springs, a mainspring which balances out a fixed and large proportion of the gravity field and a fine measuring spring which responds sensitively to the remaining small and varying fraction of the field.

In practice gravimeters do not measure deflections. It is more satisfactory to return the beam to a null position and measure the force required to do so. The usual way of doing this is to attach the upper end of the mainspring to a micrometer head and measure the displacement needed to restore the beam to its null position in terms of the micrometer reading.

Temperature compensating devices have to be incorporated in all gravimeter spring systems to make extension as independent as possible of temperature change. Even so, to achieve maximum accuracy it is necessary to keep the instrument at a constant temperature by means of a small battery operated thermostatically controlled heater. Modern gravimeters have a world wide range and standard instruments can be read to an accuracy of 0.1 g.u. New high precision instruments are also now available which can be read to 0.01 g.u.

All spring systems show a slow creep and so even if a gravimeter is kept at one place the reading will change with time. This slow change in reading is called "drift". Such instruments therefore can be used only to measure differences in gravity between locations. Though the drift rate may vary a little from day to day and even during the course of a single day the amount of drift is approximately proportional to time over a limited period. The rate may be found by returning to a base station after making a series of measurements. Provided the times were noted at which

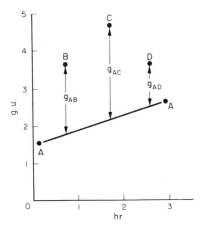

Fig. 7.3   The drift correction: g_{AB}, g_{AC}, and g_{AD} are the gravity differences between the base station A and the stations B, C and D.

the measurements were made a drift correction at each station may be made by linear interpolation as shown in Fig. 7.3. Drift rates for different instruments vary widely from 10 g.u. per day to 10 g.u. per month or less and depend to some extent on the conditions under which the instrument is being used. Provided that the returns to base are not more than a few hours apart tidal changes in gravity can be considered linear and treated as part of the drift correction.

Instruments in common use are the Worden and the La Coste and Romberg gravimeters. Both are portable and extremely sensitive instruments. With thermostat battery and case the La Coste meter weighs less than 10 kg. It has an accuracy over the full range of 70,000 g.u. of a few tenths of a gravity unit and drift, which is not sensitive to vibration, is usually less than 5 g.u. per month. Though the drift properties of the Worden are not as good, it also has high sensitivity and is lighter and more portable.

Standard gravimeters have been adapted for shallow under-water work. The instrument is housed in a watertight container lowered to the sea or lake bed and there levelled, either automatically or by remote control. An electrical signal proportional to scale reading is passed up the control cable to enable the instrument to be read at the surface.

Measuring gravity on board a moving surface ship is much more difficult since the instrument is subjected to accelerations due to wave motion. Also significant is the Eötvos effect, caused by changes in centripetal acceleration brought about by the east–west component of the ship's velocity. Since even at latitude 60° an east–west velocity change of one kilometre an hour makes a difference of about 60 g.u., it is necessary to know the ship's speed and heading very accurately. The LaCoste gravimeter has been very successfully adapted to work on a ship by mounting it on a gyro-stabilized platform. The instrument is kept vertical on its platform and so is affected only by the vertical component of the wave accelerations. These accelerations are more or less periodic so their effect on the gravimeter can be filtered out by taking a mean of the readings over a period of a few minutes. Ship-borne gravimeters are always linked to a pen recorder, filtering and various corrections being applied automatically. Ship size, sea state and quality of navigation mainly determine the accuracy of the data, agreement of repeated measurements to 10 g.u. being considered good.

There are various ways of calibrating gravimeters; for example this is usually done by the makers by means of a tilt table. This is a plane surface which can be rotated through a small angular range about a horizontal axis. The gravimeter is set up on the table and the deflection read for various small angles of tilt. Since the gravimeter responds only to the component of gravity along its axis it is acted on only by a force $g_a = g \cos \theta$. Therefore, the gravity change is $g = g(1 - \cos \theta)$. To obtain a satisfactory calibration the angle of tilt has to be known with extreme accuracy.

An alternative method of calibration is to make several determinations of the change of scale reading when the meter is run between two points of accurately known gravity difference. The essentials of a satisfactory "baseline" are a relatively short length, which limits errors due to drift, combined with a large gravity difference.

7.4 The Survey

The procedure for carrying out a survey depends a great deal on its purpose. Large scale surveys, covering hundreds of square kilometres and carried out for the purpose of revealing major geological structures, are done by vehicle or helicopter with a density of perhaps one station per square kilometre or less. The stations are conveniently placed so as to give reasonably uniform cover and may be at the points of a grid if the terrain is flat and open.

In carrying out such a survey a number of widely separated base stations are established in the area first, starting from the reference point. To achieve a high accuracy the base stations are measured in a series of closed triangular or polygonal traverses, each side of which is measured at least twice. One way of doing this is to measure a triangle of three stations in the order ABABCBCACA. In the series two drift corrected semi-independent values of each of the differences AB, BC and CA are obtained. If there were no observational errors the algebraic sum of the differences round the triangle would be zero. The actual sum is called the closing error. Further triangles, e.g. BCD etc., are now similarly measured until the base network has been completed, after which the closure errors throughout the whole network are distributed by a least squares method. This process adjusts the various differences so that the resulting values are on average the best approximations to the true differences that can be obtained with the given data. Secondary stations are now put in, starting and finishing a set at a previously measured base station so that a drift correction can be made. The observations are then corrected as already explained and a contour map prepared showing lines of equal Bouguer anomaly. Provided all reductions have been correctly made the map should reflect only gravity changes due to subsurface geological structure. If the survey only covers a small area it may not be necessary to measure a base station network first, an occasional return to one reference point to check for drift being all that is required.

Since station positions and heights have to be known accurately a gravity survey requires a considerable amount of preliminary topographic survey. In settled areas much of this has usually been done already, but in remote areas the map has to be made before the survey can be carried out. Some geological reconnaissance must also be done, together with the collection of unweathered rock samples, so that data on rock density can be made available for the elevation and terrain corrections, and later for the interpretation of the survey.

When carrying out surveys in oceanic areas position fixing becomes of the greatest importance. Long range radio navigational systems such as Omega and systems by which the ship's position is fixed relative to satellites with precisely known orbits have to be used, not only to enable the traverse positions to be recovered but to make it possible to apply latitude and Eötvos corrections with adequate accuracy. Sea depths are recorded continuously with a precision echo-sounder so that the Bouguer correction can be made. This is necessary in order to remove the variations in gravity brought about by changes in sea depth. At each "station" on a survey line the sea is replaced by rock using the simple slab formula and a density equal to that of the seabed minus 1.03 t m^{-3}, the density of sea water.

7.5 Interpretation

Bouguer anomaly maps look very like topographic contour maps. They show circular, elongated, and irregular areas of high and low gravity. They may also show linear belts of steep gradients which are not necessarily associated with any of the features just mentioned. It is possible merely from inspection of the map to make a tentative and qualitative interpretation if something is known about the geology. Gravity highs are in many areas associated with anticlines or with uplifted blocks, both being structures which bring older denser rocks nearer the surface. In other regions gravity highs may be due to the presence of heavy basic intrusions. Conversely sedimentary basins and relatively light acid intrusions usually produce gravity lows. The belts of marked gradient are produced by steep contacts between rocks of different density such as may occur across fault planes.

The problem is therefore to determine from the reduced observations used in the preparation of the map, and from what other information is available, the size, shape and position of the subsurface structures giving rise to the gravity disturbances. Although it is relatively easy to calculate the gravity anomaly of a body of any given shape there is no unique solution to the inverse problem of calculating the parameters of a body from its anomaly alone when no other data about it are available. However, usually something is already known about the geological structure of the area under survey, and from this knowledge and a qualitative examination of the gravity anomaly an idea of the size, shape and position of the structure producing it can usually be deduced. It is then possible to construct a range of possible solutions having gravity anomalies which fit the observed field to the required degree of accuracy, the limits of the range being governed to a considerable extent by the amount of the geological information. What we are doing by using such information is of course to reduce the number of unknown quantities. These are the dimensions and position of the structure producing the anomaly and the density contrast involved. If we have a fair idea of the relative dimensions (i.e. the shape) of the structure and the density contrast, then within limits we can produce a unique solution. For example a wide range of lens shaped bodies can be devised which, if placed at the correct depths, have gravity anomalies virtually identical in form and magnitude with that due to a sphere of some particular mass placed at a rather greater depth, and in the absence of other information no one solution is preferable to another. This is illustrated by the examples of Fig. 7.4. On the other hand if it is known that a gravity anomaly is in fact due to a roughly spherical body, then from the anomaly its depth and "excess mass" can be easily calculated. In this context by "excess mass" is meant the product of the volume of the sphere and its density contrast $\rho - \rho_o$ with the surrounding rock of density ρ_o. If this density contrast is known then the radius of the sphere can also be calculated and a complete interpretation has been made. Similarly a knowledge of the geological sequence and the density of the strata in an area often impose severe limitations on the possible interpretations and may make possible a solution which has practical value. Unfortunately the problem is complicated further by the fact that the anomaly field at the surface is not necessarily produced by one simple structure, but may be due to the sum of the effects of a number of quite unrelated structures at various depths. In such circumstances the first problem of interpretation is therefore to find a means, as far as this can be done, of separating out these different anomalies. This is only practicable if they divide into two groups, one containing the small scale so-called local anomalies, and the other the very much smoother and broader features, often but by no means always of considerably deeper origin. The composite field of the latter is known as the regional. Separation of local and regional field becomes a problem of filtering. The

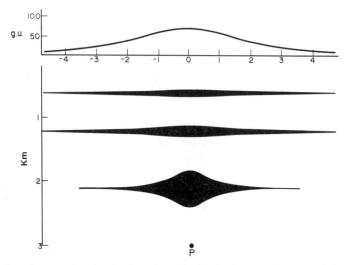

Fig. 7.4   Cross section showing lens shaped bodies having a gravity anomaly identical with that of a sphere at *P* of radius 600 m and density contrast 1.0 t m^{-3}. The thickness of the bodies is exaggerated by a factor of 3.

simplest method of separation, open to objection on grounds of subjectivity, is to smooth through the local anomalies by eye, i.e. to interpolate the regional field where it is affected by local anomalies. The smooth regional is then subtracted from the original Bouguer anomaly field, leaving a residual containing only the short "wavelength" features supposedly due to relatively small near-surface features. A more objective method of obtaining a regional is to remove the local anomalies by numerical smoothing or digital filtering (Meskó, 1965). Numerical methods are most easily used when the data are equally spaced. The simplest way of providing some smoothing is by taking a running average of groups of 3 or more evenly spaced points along a traverse. The amplitude of sharp anomalies is reduced, the amount depending very much on the overall width of the group of points compared with the anomaly wavelength. This particular averaging operation is equivalent to multiplying each of the three values by an equal weight of one-third and adding them, the central point being replaced by the sum. Improved smoothing properties are obtained if the weights decrease out from the centre and if more values are included. A five point filter might have weights of $^1/_{16}$, $^1/_4$, $^3/_8$, $^1/_4$, $^1/_{16}$. If the observations are at the points of a grid weighted averages are taken of values lying on circles of increasing radii round the centre point. The weights and the actual lengths of the radii of the averaging circles determine the behaviour of the filter. It is not difficult to design a filter with specific smoothing properties but the gravity field may be so complex that it is impossible to adequately reduce the amplitude of the local anomalies without at the same time significantly affecting the form of some of the broader features constituting the regional field. In other words a clear separation into regional and residual is not possible. Except in rather simple areas then the regional will still contain a much attenuated contribution due to the local anomalies, and on subtraction of the regional from the unfiltered Bouguer anomaly the resulting residual local anomalies will show a degree of distortion. If only qualitative interpretation is to be carried out, the aim being to enhance local trends and show up features otherwise masked by the regional, this distortion is immaterial. It may, however, considerably affect the result of any quantitative interpretation.

The trial and error method is one possible approach to quantitative interpretation. From inspection of the contour map, taking into consideration all other information about the region, a possible model of the structure is devised and its gravity effect calculated. The observed and calculated anomalies are compared, the model then progressively modified and its anomaly recomputed until a reasonable fit between the two is obtained. If the geological data are scanty it may not be possible to do more than calculate a range of approximate solutions, but even to be able to set limits to the possibilities can be most useful.

When choosing a model it is usual to make the simplest of geometrical approximations as the gravitational attraction of a number of simple forms can be easily calculated from graphs and standard formulae. Since the postulated geometrical shape can at best be only a crude approximation to the real structure the fit between the calculated and observed anomalies, though often surprisingly good, is unlikely to be "perfect", i.e. within the limits of observational error, but the procedure does enable some estimate of dimensions and depth to be made fairly simply. In many instances where there is little geological control there may well be little point in pursuing the interpretation further, since it is highly unlikely that a more complicated model will be a better approximation to the true structure, however exact the fit between the observed and calculated anomalies.

7.51 *Characteristic curves*

A more direct approach is sometimes possible. In this use is made of what are known as "characteristic curves". Changes in the parameters associated with a body (e.g. its dimensions, depth, density, etc.) are reflected in changes in the size and shape of the anomaly. Some of these are very obvious, some quite subtle. Take as an example the anomaly of the sphere shown in Fig. 7.5. It is found that the horizontal distance between the points of maximum and half maximum amplitude (the "half width") is directly proportional to the depth of the sphere. In fact the depth $z = 1.305w_{1/2}$ where $w_{1/2}$ is the half width. We could therefore (though it is not necessary) draw a straight line graph relating half width to depth to centre for the sphere. This would be a very simple "characteristic curve".

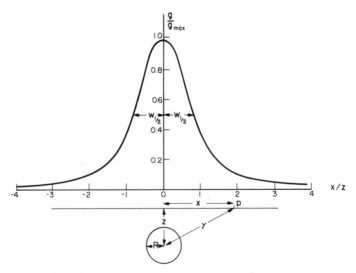

Fig. 7.5   The gravity anomaly of a sphere. Distances are given as a multiple of the depth and the anomaly as a fraction of its maximum value g_{max}.

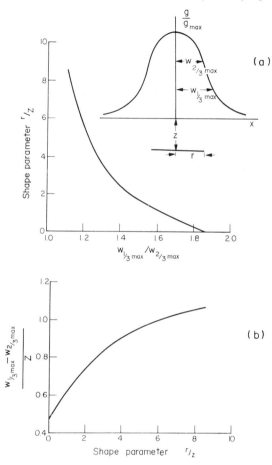

Fig. 7.6 (a) The gravity anomaly of a thin horizontal disc. (b) Characteristic curves for determining (i) the ratio of radius to depth and (ii) the depth to the disc.

In the example given the only variable that affects the half width is the depth. If as shown in Fig. 7.6(a) the body is a thin horizontal disc both depth and radius of the disc will affect the half width. In terms of the disc model a solution for depth and radius can be obtained by taking two measurements from the curve. In this instance the widths of the anomaly from the centre to one-third and two-thirds of its maximum value have been used. The variation in the ratio of these two parameters with change in the ratio of disc radius to depth was calculated, these data being shown graphically in Fig. 7.6(b). A second complementary curve is used to find the depth. From this, knowing r/z, the quantity $(w_{\frac{2}{3}} - w_{\frac{1}{3}})/z$ is found. As $(w_{\frac{2}{3}} - w_{\frac{1}{3}})$ can be obtained by measurement off the anomaly the depth of the disc may be determined and so also the radius.

More complicated characteristic curves have been calculated for structures such as inclined slabs. A study of calculated anomalies shows that the dip of the slab is reflected in the asymmetry of the anomaly, a good measure of this asymmetry being the ratio of the maximum slopes of the anomaly on the two flanks as shown in Fig. 7.7. Variations in the depth/slab length ratio are reflected in the sharpness of the anomaly, most simply measured, as described above for the disc, by the ratio of the widths at the ⅓-maximum and ⅔-maximum anomaly. Depths to top are calculated by multiplying some measure of anomaly width by a depth factor appropriate for the inclination and depth/length ratio. Characteristic curves have been calculated and published for some other simple

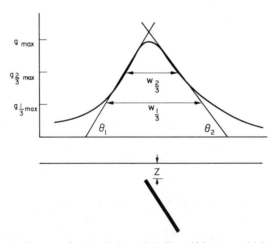

Fig. 7.7   The gravity anomaly of an inclined slab. The widths at one-third and two-thirds maximum anomaly and the maximum slopes on the two flanks θ_1 and θ_2 are parameters that may be used to determine the depth and inclination of the slab.

structures but their preparation involves a considerable amount of thought and computation, and so the number of such curves is limited. Unfortunately the simple models for which it is practicable to prepare such curves are not necessarily good approximations to the geological structures they are being used to represent and so apparently acceptable solutions can be considerably in error. Thus, although such curves can be very helpful their application should be restricted to situations in which the gravity data are detailed and accurate and where enough is known about the geology to ensure that a realistic model is being used.

Spheres, cylinders and slabs are obvious shapes to choose when making preliminary interpretations. A few examples are given below as illustrations. Standard references (see Bibliography) should be consulted for a fuller treatment.

7.52 The sphere

A spherical mass of radius R lies with its centre at a depth z below the surface. If its density were the same as its surroundings it would produce no gravity anomaly. Therefore in calculating its anomaly the density difference from the country rock must be used. Suppose this is equal to σ. Then the excess mass is $4\pi R^3 \sigma /3$. As the mass of a sphere can be considered to act from its centre the field due to this excess mass at a point at the surface distance r from the centre of the sphere will be

$$G \cdot \frac{4\pi R^3 \sigma}{3r^2} = G \cdot \frac{4\pi R^3 \sigma}{3(x^2 + z^2)}$$

since from Fig. 7.5 $r^2 = x^2 + z^2$, The vertical component of this attraction is

$$g_z = \frac{4G\pi R^3 \sigma}{3} \cdot \frac{z}{(z^2 + x^2)^{3/2}}$$

since $\cos \theta = z/r$.

For ease of use it is convenient to express the horizontal distance in terms of units of depth. The expression is rearranged as follows:

$$g_z = \frac{4G\pi R^3 \sigma}{3z^2} \cdot \frac{1}{(1 + x^2/z^2)^{3/2}} \tag{7.10}$$

The function $(1 + x^2/z^2)^{-3/2}$ was used to plot Fig. 7.5 with x/z as the abscissa. It is only necessary to multiply the values on the graph by $4G\pi R^3\sigma/3z^2$ to obtain a complete profile over the chosen sphere. Care must be taken that the value of G used is consistent with the units in which dimensions and density contrasts are measured: it is convenient to remember that for dimensions in metres, densities in t m^{-3} and gravity field in g.u., the factor $4\pi G$ is twice that appearing in the Bouguer correction and is 0.84 g.u. t^{-1} m^2.

As previously discussed the depth to the centre of a spherical body is related to the "half width" by the expression $z = 1.305x_{1/2}$. Having measured the half width of an observed anomaly thought to be due to a roughly spherical body, and thus obtained a depth to centre, the excess mass $M = 4\pi R^3\sigma/3$ can then easily be obtained by substituting the value for the observed maximum anomaly into the formula, remembering that the maximum occurs at $x = 0$. Thus $g_{max} = GM/z^2$.

7.53 *The cylinder*

Geological structures in areas of sedimentary rocks are more often than not elongated in the strike direction and quite often roughly horizontal. Steep anticlines may be simulated by horizontal cylinders, these being assumed to be of infinite length in order to simplify the expression to be used in calculating the anomalies. For an infinite buried cylinder where the excess mass per unit length $M_1 = \pi R^2\sigma$

$$g_z = \frac{2G\pi R^2\sigma}{z} \cdot \frac{1}{(1 + x^2/z^2)} \tag{7.11}$$

In this instance the depth to the centre is equal to the "half width" defined as for the sphere, i.e. the depth factor is unity rather than 1.305.

7.54 *The slab*

Where a vertical fault occurs in a simple sequence consisting only of two horizontal beds A and B, the lower being considered for practical purposes infinitely thick, the gravity effect can be represented by that of a horizontal slab with a density equal to the difference between those of A

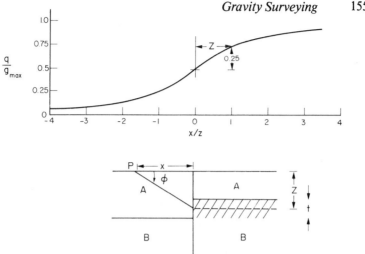

Fig. 7.8 The gravity anomaly across a vertical fault displacing a horizontal interface. The lower medium B has the higher density.

and B. In Fig. 7.8 the shaded area shows the form of the slab. This is the volume in which medium B has replaced A and is the only region effective in causing an anomaly. Provided the slab is relatively thin compared with its depth a very simple approximate formula can be used to compute the anomaly. This assumes all the excess mass to be condensed on the median plane of the slab. This excess mass is therefore σt per unit area of the sheet where σ is the density contrast and t the thickness of the sheet. The gravity effect can be shown to be $G\sigma t$ multiplied by the solid angle subtended by the sheet at the point of observation. For a horizontal semi-infinite sheet, using the notation of the figure this solid angle is 2ϕ which is twice the angle measured in the plane of the diagram. Therefore

$$g_z = 2G\sigma t\phi$$

$$= 2G\sigma t\left(\frac{\pi}{2} - \tan^{-1}\frac{x}{z}\right) \qquad (7.12)$$

The form of the anomaly is given in the figure.

The total change of gravity across the fault $2\pi G\sigma\, t$ is of course equal to the Bouguer anomaly of an infinite slab of thickness t and density contrast σ, and over the edge of the fault the change is half the total.

The depth to the centre plane of the slab is equal to the distance from the fault to the point at which the gravity change is one-quarter of the total range of the anomaly.

The formula can easily be used to calculate the gravity effect of rectangular blocks infinite in a horizontal direction perpendicular to the traverse provided they are not too thick. If the angle subtended by the median plane of the block is θ the solid angle is 2θ and the gravity effect is

$$g_z = 2G\sigma\, t\theta \qquad (7.13)$$

If the actual length perpendicular to the traverse is three or more times the distance to the point at which the anomaly is being calculated the error in assuming infinite length is quite small and can be neglected for most practical purposes.

7.55 Irregular bodies

Though the gravity effect of bodies of irregular shape is usually calculated by computer the procedures are relatively straightforward if the bodies have one horizontal dimension long enough to be considered infinite for computational purposes and a more or less constant cross section. Such bodies are known as "two-dimensional". The total anomaly is calculated by using the formula for gravity effect at the surface of an infinite horizontal rod-like element of small area and integrating numerically over the whole cross section of the body. This can also be done manually using a graticule, a transparent sheet divided into compartments by concentric circles and radii. The cross section of the body is drawn out on a suitable scale and the centre of the chart placed at the station. The sizes of the compartments are so arranged that the gravity effect at the origin of the mass contained within each one is the same. Thus it is only necessary to count the number of compartments contained within the body cross section and to multiply by the effect of each to obtain the anomaly at a particular point. A simple graticule is illustrated in Fig. 7.9.

O   A

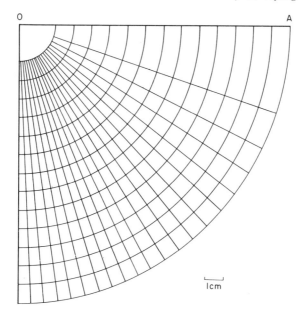

Icm

Fig. 7.9   A graticule for the calculation of the gravity anomalies of "two-dimensional" bodies. The effect of each compartment is 6.67×10^{-5} σS gravity units where the scale is $1:S$ and σ is the density contrast in t m^{-3}.

The gravity effect of one compartment is given by

$$g = 2G\sigma(r_{m+1} - r_m)(\cos\phi_{n+1} - \cos\phi_n) \qquad (7.14)$$

Computational methods of various kinds have been devised for obtaining the effects of two-dimensional bodies of irregular shape. One way is to approximate the cross section of the body by an irregular polygon, the number of sides being made large enough for the error involved to be small (Talwani *et al.,* 1959). Methods are also available for computing the anomalies of irregularly shaped bodies which are limited in all three dimensions. The surface of the body may, for example, be replaced by a set of polygonal faces (Okabe, 1979), making it possible to specify the body position and dimensions with a limited number of coordinates, thus simplifying the calculation. As used in interpretation these methods are "indirect" in that the body parameters are specified, the anomaly is computed and then compared with observation. An important recent development is towards the rapid computation of the parameters of a body of specified shape directly from a knowledge of the gravity anomaly.

It is worth restating in this context that whatever the method used a unique solution is not possible unless some geological information is available to provide constraints. Also a "unique" solution is in itself likely to be unique only in the sense of providing an answer within certain limits of error, for resolution in gravity interpretation is not good.

7.6 Examples

We shall conclude this chapter with a brief discussion of the interpretation of two gravity surveys which together illustrate many of the points that have been discussed.

The first is described by Bott and Masson-Smith (1960) and was undertaken to determine the subsurface form of the Criffel granodiorite mass near Dumfries on the Solway Firth. The map of Bouguer anomalies (Fig. 7.10) shows a roughly circular gravity "low", with another and more elongated low lying a few miles to the north-east. The latter is clearly due to the basin of light Permo-Triassic sandstone round Dumfries and will not be further considered here except to point out that, although there is an area of overlap between the anomalies, the problem of isolating them, which is the necessary preliminary to interpretation, is not in this case a difficult one. It is also an easy matter to decide that the circular low is in fact almost certainly due to the granodiorite mass itself. The gradients of the anomaly are so steep that its source cannot lie deeper than about 8 km. This implies an origin in the upper crust and it seems likely that the granodiorite, which was shown by direct measurement to be of low density compared with the country rock, is responsible. Surface density measurements also showed that a region of transitional density exists on the south-east margin of the intrusion. The first stage of the interpretation was to assume that the mass was of a simple form consistent with these surface outcrops, but with an infinite length perpendicular to the profile AA', and to adjust the slope of the

Fig. 7.10 Bouguer anomalies near Dumfries (Bott and Masson-Smith, *Proc. Yorks. geol. Soc.,* 1960).

boundaries of this "two-dimensional" model and the position of the base of the intrusion until its computed anomaly was in the best possible agreement with that observed along the profile. Figure 7.11 shows the unacceptably poor fit which was obtained even with the best values of these parameters, and Fig. 7.12 shows a much better fit with the field observations obtained by supposing that the known zone of intermediate density in fact extended further to the north-west. The validity of this

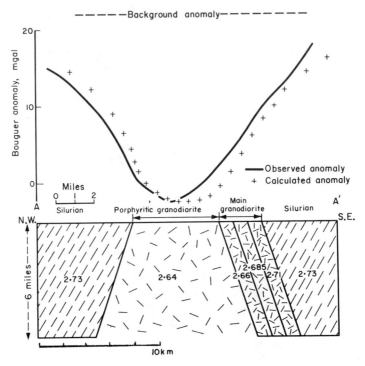

Fig. 7.11 Calculated and observed gravity anomalies over the Criffel granodiorite.

interpretation depends on the density contrasts remaining the same at depth as they are at the surface, and of course other modifications to the shape of the intrusion could have been postulated to give the improvement in fit between Figs. 7.11 and 7.12. With these reservations, the authors state that Fig. 7.12 represents the best interpretation of all the available geological and geophysical evidence: a full "three-dimensional" interpretation leads to a similar shape but a depth about 15 per cent greater.

The second example shows to what extent gravity measurements can be expected to help in determining the cross-section of an alluvium-filled valley. The map of Bouguer anomalies shown in Fig. 7.13 (Blundell,

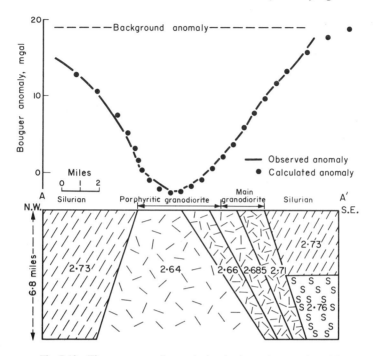

Fig. 7.12   The same anomalies recalculated using an improved model.

Griffiths and King, 1969) covers an area in North Wales between Portmadoc and Harlech and suggests that the alluvium-filled valley north of Portmadoc continues southward beneath a level sand-flat, since both the known valley and the postulated concealed one give rise to similarly elongated gravity lows. The interpretation is here confused by the presence of a regional gradient originating in structures at depth in the solid rocks and revealed by a large scale survey carried out with a wider station spacing. The regional map, which smooths out the local effect of the buried valley, has a gradient of about 12 g.u. km^{-1}, the contours trending across the valley in a roughly NW–SE direction, as can be seen on the Bouguer anomaly map in those areas away from the influence of the valley fill. The residual anomaly, obtained by subtraction of the

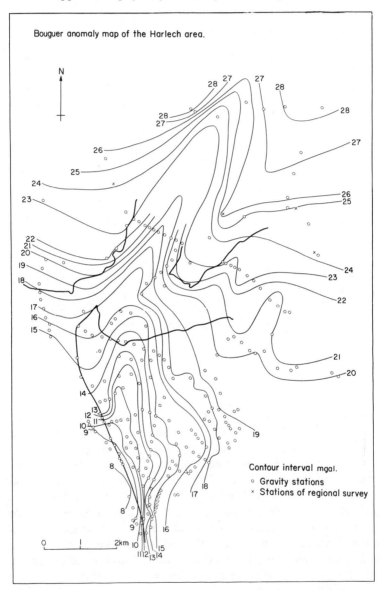

Fig. 7.13   Bouguer anomalies between Portmadoc and Harlech.

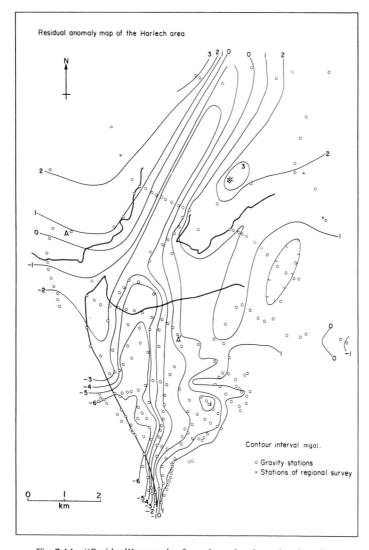

Residual anomaly map of the Harlech area.

Contour interval mgal.

○ Gravity stations
× Stations of regional survey

Fig. 7.14   "Residual" anomaly after subtracting the regional gradient.

regional values, is shown in Fig. 7.14. The zero contour fairly closely follows the boundary of the alluvial fill and over the solid rock outcrop the residual anomaly has only small values, which suggests that the observed anomaly has in fact been separated into the two parts required; one due to the light alluvial fill of the valley, the other to deeper density contrasts in the solid rock which are not the concern of this survey.

In this kind of problem the surface geological control is generally good in that the upper surface and lateral extent of the alluvial mass causing the anomaly are well defined, but as in the first example it has to be assumed that the densities of hard rock and alluvium measured on the surface do not change appreciably either with depth or laterally. In the particular example illustrated there was the further difficulty that the alluvium was too friable for reliable density measurements on samples to be made. It was thus possible, by trial and error methods, to determine the form of the valley which best fitted the gravity observations, but not to put a vertical scale on the section since the density contrast was unknown. A section across the line AA', with its computed anomaly, is given in Fig. 7.15: changes in the detailed form of the section can be made without significantly altering the gravity anomaly. No borehole was available to give the required control, but a seismic refraction profile across the same valley did give a less ambiguous estimate of its maximum depth as 240 m, which implied a density contrast of 0.7 t m^{-3}, i.e. a density of about 2.0 t m^{-3} for the alluvial material. This density was then used to determine the form of the valley over a much wider area than that covered by the seismic profile. Unfortunately an element of ambiguity remained in the interpretation of the seismic profile itself, since there was some evidence that a "hidden layer" (see §3.53) of intermediate velocity, not thick enough to give first arrivals on the seismograms, existed in the alluvial layer. If this zone existed, the maximum depth h estimated from intercept times t_i (eqns. 2.7, 2.8) would be increased with the increased average value of the velocity of the alluvium to about 360 m and the assumed density contrast would have to be reduced to about 0.5 t m^{-3} if the gravity anomaly were to be consistent with this greater depth. This means that the average density of the alluvium would have to be 2.2 t m^{-3} rather than 2.0 t m^{-3}, which is quite consistent with the presence of a masked layer of higher velocity and, therefore, probably higher density. In this example then, even a combination of two geophysical methods does not give a unique result, but of course a single suitably placed

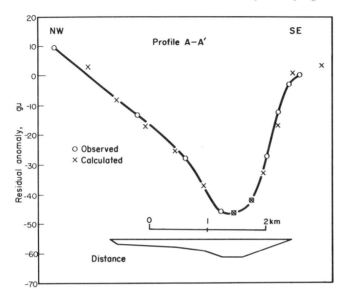

Fig. 7.15 A possible cross-section of the Portmadoc-Harlech buried valley. A density contrast of 0.7 t m^{-3} is assumed.

borehole would provide the key to both the seismic and the gravity interpretations and would enable the form of the valley to be plotted with considerable accuracy over the whole area of the gravity survey.

Surveys of this type, in which the form of a valley or basin filled with light sediments is determined, have obvious applications in the initial stages of prospecting for petroleum, though seismic methods are required to define the traps present. On a smaller scale, alluvium-filled valleys may be of interest as aquifers, but although a gravity survey may provide a cheap way of determining their extent and thickness, this thickness must exceed 30 m or so if a useful anomaly of 10 g.u. or more is needed, since the density contrast between alluvium and bedrock is unlikely to exceed about 0.8 t m^{-3}. Smaller density contrasts would of course imply a greater minimum measurable thickness. Gravity surveys can also be used for targets such as ore-bodies, or voids in site investigations, but even though the contrasts are here of the order of $2\text{–}3 \text{ t m}^{-3}$, success is unlikely if the depth to the top of the body or void is much greater than its diameter, or if there are marked variations in overburden or topography.

CHAPTER 8

Magnetic Surveying

8.1 Introduction

THOUGH there are certain marked similarities between magnetic and gravitational methods, both in the field techniques and the presentation of data there are important differences.

Gravity surveys depend for their effectiveness on subsurface density differences and magnetic surveys make use of the variation in magnetization of rocks. Density is a bulk property of rocks and tends to be consistent throughout a formation. Rocks, however, owe their magnetic properties to minor constituents which can be very variable in their distribution, and thus less diagnostic of the formation.

In igneous and metamorphic rocks the main carriers of the magnetization are the mixed oxides of iron and titanium, of which the iron rich member magnetite is the best known. Sediments, in general less magnetic, contain magnetic particles derived originally from igneous and metamorphic rocks and also oxidation products, haematite being particularly important. The iron sulphide pyrrhotite contributes significantly to the magnetization in a limited number of rock types, occurring for example in association with the valuable nickel ore pentlandite.

As a result of the presence of the earth's field rocks containing magnetizable minerals show an induced magnetization. The constant of proportionality between the inducing field and the magnetization is known as the susceptibility. The inducing field, however, is the field within the magnetized material, though in rocks, which can normally be classed as weakly magnetic bodies, this does not differ much from the earth's field. It is therefore convenient to define an apparent susceptibility k relating the magnetization to the undisturbed applied field. This relationship has the form

$$\mu_o J = kB \tag{8.1}$$

where J is the induced magnetization (in A m^{-1}), B the field strength measured in teslas (T) and μ_o a constant with dimensions of H m^{-1} (see §1.2). It holds for most rocks (i.e. magnetic content < 10 per cent) but for very magnetic bodies the value of k will depend on body shape and field direction. With the exception of certain strongly magnetized bodies the induced magnetization lies in the direction of the inducing field. Rocks may also show a remanent or permanent magnetization. This can be acquired in a number of different ways. Igneous rocks become magnetized on cooling in the earth's field and so this phase of the magnetization may date from the time of their formation if it is sufficiently permanent or "hard". Sedimentary rocks may acquire a remanence during deposition in water, this being due to the alignment of magnetic particles on settling in the earth's field. Chemical changes during diagenesis may also produce a remanence. Whatever the cause the direction of the remanence in rocks is, with some exceptions, close to that of the geomagnetic field at the time of formation of the magnetization. Its direction today may be very different, for there may have been changes in the position of the magnetic poles relative to the rock mass, and also faulting and folding of the formation. The stability of this remanence in both magnitude and direction depends on both the mineralogy and the texture of the magnetic minerals. In some rocks, as a result of age and instability of the magnetic minerals, its magnitude may be insignificant (see Tarling, 1971).

Induced and remanent magnetism add vectorially to give a resultant, the direction of which depends on the strength and direction of the remanence relative to that of the induced component.

In considering local anomalies in the earth's magnetic field we have to take into account not only the magnetization of the rock but the important fact, later discussed, that for a magnetic body of a given shape the form of the anomaly is governed not only by this shape but also by the inclination of the present direction of the earth's field and by the body orientation. We are thus faced, in dealing with magnetic anomalies, with a far more variable and thus less diagnostic feature than we have to deal

Table 8.1   Magnetic properties of some common rock types.

	K, SI x 10^6					
	I 10 10^2 10^3 10^4 10^5 10^6					
Basic volcanics						
Basic plutonics						
Granites						
Gneiss, Schist, Slate						
Sedimentary rocks						

with in gravity interpretation. Though it is certainly possible to make broad distinctions between weakly and strongly magnetized rocks (prominent anomalies in fact most often arise as a result of the juxtaposition of the members of the two types) interpretation in terms of body size and shape is far less easy. For this reason much magnetic survey tends to be of a reconnaissance nature and interpretation is qualitative or at best semi-quantitative.

8.2 The Earth's Magnetic Field

The reduction of magnetic field data for the purpose of producing an anomaly map generally requires a knowledge of the normal variation of the geomagnetic field in space and time in the area of the survey. A brief discussion of the relevance of this subject to prospecting is therefore given here.

To a rough approximation the form of the magnetic field at the earth's surface is that which would be produced by a small but powerful magnet placed at the earth's centre with its north magnetic pole pointing southwards and inclined at 11° to the rotation axis (Fig. 8.1). If the field were perfectly regular the lines of force would be vertical at the pole of the magnetic axis and horizontal on the magnetic equator, this being a great circle inclined at 11° to the true equator. All isodynamic lines, or lines of equal force, would be small circles round the earth parallel to the

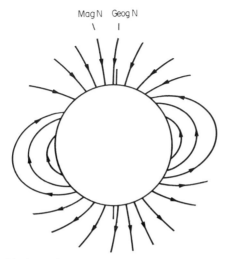

Fig. 8.1 The form of the earth's magnetic field.

magnetic equator. In fact the actual pattern of the field is only roughly of this form as there are regions of up to continental size where there are marked departures from this simple picture. One result of the added complexity is that the dip poles, where the inclination at the surface is 90°, do not lie on the magnetic axis. In addition both the strength and form of the field change slowly by amounts which are quite easily measured over a period of a year. Despite the complex form of the field and its variation with time it can be described mathematically with adequate accuracy, though the expression used contains a large number of terms. The field so calculated is known as the International Geomagnetic Reference Field (1976). At the poles the strength of the field is about 6×10^{-5} T. Magnetic anomalies have amplitudes that are only a small fraction of this and are measured in nanoteslas (1 nT = 10^{-9} T), more commonly known as gammas (γ).

Figure 8.2 shows the conventions and nomenclature adopted in describing the field. The total field vector is denoted by F, H and Z being the horizontal and vertical components. The angle of inclination of the field is I and D, the angle the horizontal component makes with the meridian, is the declination.

Applied Geophysics for Geologists and Engineers

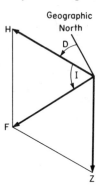

Fig. 8.2   The elements of the earth's magnetic field.

As well as the slow secular change of the field already referred to it is subject also to short periodic variations. There is a roughly cyclic change with a 24-hour period known as the diurnal variation. The amplitude of the daily variation is dependent on latitude and is not altogether constant from day to day, some days being more disturbed than others. In mid-latitudes the variation on quiet days is some tens of gammas. There are also sudden disturbances of the field known as magnetic storms which show a basically regular pattern on which are superimposed irregular changes. Large storms produce changes of as much as several hundred gammas and the initial disturbance takes several days to die away. The daily variation is caused by electric currents due to tidal movements and ionization in the ionosphere. Magnetic storms are caused by the circulation round the earth of charged particles from the sun in a region beyond the atmosphere known as the magnetosphere.

8.3 Instruments

Virtually all magnetic prospecting today is carried out with electronic instruments, rather than those of the mechanical balance type. Apart from the high accuracy obtainable and the ease with which a record of the field may be obtained such instruments do not require a stable platform and thus can be used at sea or in the air as well as on land.

Instruments using the *fluxgate* principle provide a continuous record of the change in the component of the field along the sensitive axis to an accuracy of ± 1 nT but they are subject to drift and therefore cannot be used to measure the absolute value of the field. To measure the total field it is necessary to maintain the axis of the fluxgate to within about a quarter of a degree of the field direction, this being done by suspending it in gimbals or in the case of an airborne or marine survey by servo-motors operated by a pair of auxiliary fluxgates mounted at right angles to the measuring element. The servo-motors operate until there is no output from the auxiliary fluxgates, when these must be perpendicular to the field and the measuring head therefore parallel to it.

The *proton magnetometer* has a number of distinct advantages over instruments of the fluxgate type. It is an absolute instrument and can be designed to give a direct reading of field strength. There are no calibration problems and inherently it is drift free. Exact orientation is not necessary since the total field is measured rather than any component. The principle of the instrument is as follows.

In a sample of a suitable liquid, such as a 500 ml container of water, the proton spins line up parallel or antiparallel to the earth's field with a small excess in the field direction. The liquid therefore has a small magnetic moment in this direction. If then a strong field is applied in a direction approximately at right angles to the earth's field the protons will be realigned to give a stronger moment in this new direction. On cutting off this field abruptly this magnetic moment will relax to the earth's field direction in a few seconds by precessing round it, the frequency of precession f (about 2 kHz) being related to the field strength by the equation

$$f = \gamma_p B / 2\pi \mu_o \qquad (8.2)$$

where γ_p = a known constant, the gyromagnetic ratio of the proton.

During the period of relaxation the protons will induce a small e.m.f. in a coil wound round the bottle and this e.m.f. will have the same frequency as that of the precession. The absolute strength of the total field therefore can be determined from an accurate measurement of this frequency.

Accurate setting up in a particular direction is not necessary but older instruments employing a simple solenoidal energizing coil should be orientated with the coil axis at a large angle to the earth's field. Newer instruments have toroidal coils and are not sensitive to orientation.

Figure 8.3 is a block diagram illustrating the working of one current version of the instrument. The precession signal is amplified and compared in phase with a high frequency oscillator. A control signal which depends on the phase difference is used to lock the oscillator frequency to a multiple of that of the precession signal. The oscillator frequency is then measured by the conventional method of counting the number of cycles in a fixed time interval, the length of which can be so chosen that the count displayed is equal to the field strength in nT, to a precision of 1 part in 50,000 rather than the 1 part in 2000 which could be achieved by counting the 2 kHz precession signal directly. Although such an instrument appears to give a continuous reading, it is of course reset only during the proton precession, which takes place at intervals of about

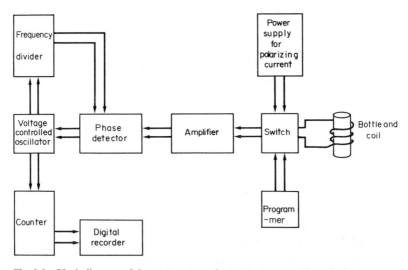

Fig. 8.3   Block diagram of the proton precession magnetometer. The polarizing current to the bottle is switched off by the programmer, after which the precession signal is fed to the phase detector. The oscillator signal is frequency divided to approximately the precession frequency (~2000 Hz) and compared with this in the phase detector. A control signal proportional to the phase difference between the two changes the oscillator frequency until it locks to the chosen multiple of the precession frequency. This is then counted.

one second, since time has to be allowed for the proton alignment by the polarizing field.

Instruments are available measuring in the range 20,000–100,000 nT. Cycling is automatic, typically giving a reading to an accuracy of 1 nT once every second, though somewhat faster repetition is possible.

In field instruments the detecting head comprising coil and bottle is set up a few metres from the electronics to avoid any interference. At sea the detector has to be towed several hundred metres astern of the ship to reduce the influence of the ship's own magnetic field to negligible proportions. When making measurements in the air the detecting head is also often towed behind the aircraft but it is possible to install it in a housing projecting from the tail. If this is done compensating fields have to be applied to annul the magnetization of the aircraft.

8.4 Magnetic Surveys

Most ground surveys are made with a proton magnetometer though portable fluxgate type instruments may be used when high sensitivity is not required or when strictly continuous field readings are required as in some airborne surveys. As already explained proton magnetometers measure the strength of the total field. In the presence of a local field with an arbitrary direction due to a magnetized body the instrument thus measures the resultant of this and the geomagnetic field. Since the latter is almost always very much larger than the anomalous field the direction of the resultant differs little from that of the undisturbed field. Hence the magnitude difference of the two is a very close approximation to the component of the anomaly in the geomagnetic field direction. This point is illustrated in the diagram of Fig. 8.4.

Most ground magnetic surveys cover only relatively small areas. Measurements are normally made at regular intervals along traverses or in a grid pattern should it be thought necessary to produce a contour map. If only profiles are being measured it is important to orientate these perpendicular to the strike of the magnetic structures if known. Station density is determined by the "wavelength" of the anomalies being measured, it being necessary to obtain enough readings (about four per wavelength) to delineate these sufficiently accurately.

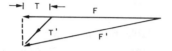

Fig. 8.4 The definition of "total field" magnetic anomaly. F= geomagnetic field, T'= anomaly field, F'= resultant T= "total field" anomaly, i.e. the component of T' in the direction of the earth's field.

Being absolute instruments proton magnetometers do not drift but periodic returns to a base during the day may be necessary to keep a check on diurnal variation. These base readings are plotted against time, the smooth curve through the points being a fair representation of diurnal change if the day is not stormy. Provided the times of all station measurements have been noted the diurnal plot can then be used to make the necessary station corrections. Such a procedure is only required where an accuracy of a few nT is aimed at, and this will normally be in areas where the anomalies are, say, 100 nT or considerably less. Where anomalies are very large and the survey is only over a small area diurnal correction can be dispensed with.

If very high accuracy is required the best way is to monitor the field at base with a second recording magnetometer. Alternatively, if there is a permanent magnetic observatory in the region copies of the magnetograms can be obtained and used for this purpose. Caution should be observed if this is done as the form and magnitude of the diurnal variation changes with distance in some areas in a somewhat complex manner.

Compared with gravity the accuracy of magnetic measurements is at least two orders of magnitude less: commonly 1 part in 10^5 rather than 1 in 10^7. Elevation and terrain corrections are therefore insignificant. In local surveys corrections for changes in the main field with position are also often too small to be of importance and even if appreciable may be extracted as a "regional".

Magnetic surveys of extensive regions are carried out using fixed wing aircraft, the economic and scientific advantages of operating in this way being very considerable. For example difficult terrain has no effect on survey conditions and many thousands of kilometres of line can be covered in a few weeks of flying. Closely spaced data (proton magnetometer with 1 s cycle time: 50 m at 180 km hr^{-1}, a typical airspeed for a light aircraft) are thus obtained quickly and have the advantage of being in the form of continuous and conveniently spaced straight and parallel profiles. Furthermore they are in a form suitable for automatic data processing and plotting.

Aeromagnetic surveys with fixed wing aircraft are, however, to some extent of a reconnaissance nature. Navigational difficulties make it very difficult to fly closely spaced lines and the smoothing of sharp anomalies due to the height of the flight lines makes close spacing pointless. Somewhat more detailed work can be done by helicopter but interesting areas may have to be followed up on the ground. In mineral exploration, where structures of interest are near surface and relatively small, anomalies tend to be of short wavelength and so lines are flown with small but constant ground clearance; this may be no more than 100 m. Flight line spacing may have to be as close as a few hundred metres to ensure no ore bodies of economic size are missed. For surveys of a more regional nature the spacing may be expanded to 1 or 2 km and the height increased to 300 m or more. When deep sedimentary basins are being sought, the source of the magnetic anomalies being the magnetic basement beneath, then a constant flight height of 1000 m and a line spacing equal to the basement depth are appropriate. Except in certain special circumstances profiles are flown across the regional geological strike, cross lines (tie lines) being spaced at an interval of perhaps ten times that of the profiles, as shown in Fig. 8.5.

Some surveys are flown using radio-navigational aids but where terrain permits a satisfactory and less expensive method is to recover flight line positions by photography. A sequence of overlapping photographs of the ground beneath the aircraft is obtained during the survey, the relationship between photographs and magnetic measurements being recorded automatically. Positions are later fixed by comparison with maps or a previously flown photomosaic of the whole area. A base magnetometer is set up to check for magnetic storms and disturbed days but the records are not always used to correct flight line data. As already mentioned the magnitude and phase of the daily variation change with distance in a somewhat uncertain way, and there is in addition a height effect. Corrections are not likely to be accurate enough over distances much exceeding 100 km from the base instrument, particularly in high magnetic

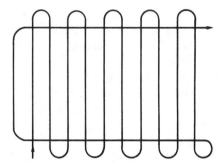

Fig. 8.5   Flight pattern for an aeromagnetic survey showing tie lines.

latitudes. Tie lines, however, provide a method of correcting for time variations in the field which is usually satisfactory. On a reasonably undisturbed day the field can be considered not to have changed during the time taken to fly a single tie line. Provided survey and tie lines are flown at the same height within close limits any differences in the measured field at crossover points can be attributed to diurnal change.

8.5 Interpretation

The approach to interpretation is controlled by many factors, of which the nature of the geology and the scale of the survey are not the least important.

Interpretation of an aeromagnetic survey of a large region of mainly unmapped territory in which the magnetic rocks are at or near surface is likely to be mainly qualitative. By inspection and using experience the interpreter will seek to uncover the main features of the geology, to outline the various rock masses and reveal as much structural information as possible. The more geological ground control there is available the more useful the magnetic data are likely to prove.

Where the magnetic formations are overlain by non-magnetic sediments the approach will be more quantitative for information on the thickness of the cover will almost certainly be required. In general so little will be known about the actual geology that a number of simplifying assumptions have to be made about the source of the anomalies to enable calculations to be made.

In interpreting detailed ground surveys much more geological information is likely to be available. In some instances when the control is good it may be possible to obtain a considerable amount of information about the nature of the body and its geometry, thus enabling a reasonably exact interpretation to be carried out. Often, however, a simple semi-quantitative approach is all that can be attempted. One such approach makes use of what is known as pole and line theory. The anomalies are assumed to be due to the magnetic fields of a subsurface distribution of magnetic poles, lines of poles, dipoles and lines of dipoles. Very crudely this is equivalent to replacing the magnetized body by an arrangement of magnets. Single poles cannot occur alone and so have to be envisaged as very long magnets, the far pole being too distant to have any influence. Dipoles on the other hand are thought of as short bar magnets. Suitable assemblages give lines of poles and dipoles.

The advantage of this simple method is that it is not necessary to make any assumptions about the geometry of the body or structure giving rise to the anomaly. It does not have to proceed beyond the stage of finding a satisfactory distribution and strength for a set of poles and dipoles, the resultant magnetic field of which approximates to that of the anomaly. Even an interpretation in these terms can be very useful, for it can provide at least a measure of the depth and position of the anomaly source, and it may give at least a general idea of its size and shape. Alternatively, in circumstances where there is rather more geological information about the source, a simple geometry for it may be guessed at the outset. Having decided on the magnetization direction, which may have to be taken as that of the earth's field unless the remanent magnetization is known, the general form of the pole distribution is fixed. The body position and size, and therefore the pole distribution, are then modified until a satisfactory fit to the observed anomaly is achieved.

There is little to be gained by making use of more than the simplest arrangements of poles and dipoles. A whole range of more complex distributions may all yield a better fit to the observed anomaly but they do not necessarily have a physical reality, and especially in the absence of geological control add nothing to the interpretation.

An important aspect of the processes described above is, of course, recognition and classification of anomalies merely by inspection. It is therefore helpful to consider some of the simple geometric shapes used to represent geological structures and the magnetic expression of these in terms of poles and dipoles etc., remembering that the latter depends also on magnetization direction. Some examples are discussed below.

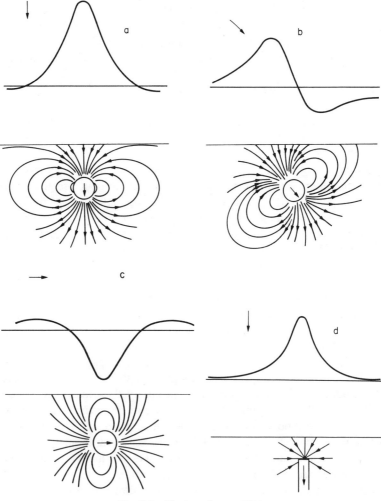

Fig. 8.6   (For legend see p. 179.)

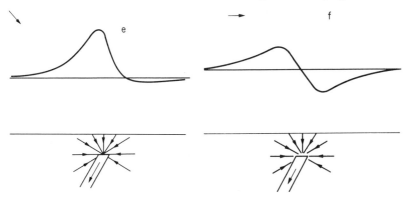

Fig.8.6 (*cont.*)   Total field anomalies for the dipole and the single pole for geomagnetic field inclinations of 90°, 45° and 0°. The dipole of Figs. 8.6(a), (b) and (c) is shown as the field of a homogeneously magnetized spherical body magnetized by induction in the direction of the earth's field. A long thin pipe with a resultant magnetization along its length has free poles only on its two end surfaces (see Fig. 8.7). If the diameter is small compared with the depth to the top surface and the pipe is long enough the anomaly approximate to that of a single pole; thus the dip of the pipe does not affect the form of the anomaly. The difference in shape of the anomalies in Figs. 8.6(d), (e) and (f) results from resolving the field of the single pole along different geomagnetic field directions.

A spherical ore body, homogeneously magnetized, has an anomaly equivalent to that of a dipole placed at its centre. Narrow vertical pipe-like bodies (i.e. vertical cylinders) magnetized along the length have poles at each end but if the body is long compared to the depth to its top the field of the lower pole may be ignored and the anomaly considered to be due only to the upper pole. Long horizontal pipe-like bodies (i.e. infinite horizontal cylinders) act as lines of magnetic dipoles on the cylinder axis, having an inclination equal to that of the component of the earth's magnetic field in a plane perpendicular to the long axis. Cross profiles look very similar to those of a dipole but are not identical with them.

Vertical dipping thin dykes (thin slabs or sheets), if magnetized in their plane, have a line of magnetic poles along the upper edge. If the field makes an angle with the plane of the sheet a cross component of magnetization will be present. The magnetization can then no longer be represented by a lines of poles, and a sheet of dipoles has to be included. Some models and their "total field" magnetic anomalies (i.e. anomaly components in the field direction) are shown in Fig. 8.6. Flux lines have been drawn in as an aid to understanding. Where the body geometry is simple, doing this is usually straightforward. From such diagrams the general form of an anomaly can often be deduced without calculation. The anomaly will be positive where the field of the body has a component in the geomagnetic field direction. Where there is an opposed component the anomaly will be negative. Flux lines at right angles to the geomagnetic field indicate points of zero total field anomaly.

Trial and error fitting of calculated to observed anomalies is a slow and tedious process even in simple cases. As in gravity interpretation, once the geometry of the body or the pole distribution giving rise to an anomaly has been guessed by inspection, then with the aid of characteristic curves an interpretation can be carried out.

Because anomaly shape varies with field inclination and many other parameters, characteristic curves for magnetic anomalies are more complicated than the corresponding curves for gravity anomalies. For example, for the simple case of a single pole in a vertical field the depth/half-width ratio has the same value as that for the gravity anomaly of a sphere since the anomalies have the same shape. The numerical value of the ratio changes, however, as the earth's field inclination alters. Graphs of such "depth factors" for poles, dipoles, lines of poles and lines of dipoles are given in a paper by Smellie (1956).

If the magnetic and geological data are sufficiently good it may be possible to make a reasonable guess as to the geometry of the body or structure giving rise to the anomaly. In such circumstances it may be worth attempting a more detailed interpretation than pole and line theory permits. Calculation proceeds on the assumption (strictly justified only for spherical or ellipsoidal bodies) of a uniform magnetization. This may be purely an induced magnetization and therefore proportional to the susceptibility and the field strength (see equation 8.1) or it may be a resultant magnetization which includes a remanent component. To evaluate the magnetic field of the body the magnetization is best expressed in terms of a distribution of "free poles" situated on the surface of the body. This is most simply illustrated by considering a cylinder of length l and cross-sectional area A, magnetized uniformly along its axis as shown in Fig. 8.7. The cylinder will have a surface of free poles at each end which may be considered to have total strength m and

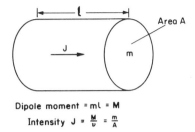

Dipole moment = m l = M

Intensity $J = \frac{M}{v} = \frac{m}{A}$

Fig. 8.7   Equivalent surface pole strengths for a cylinder magnetized along its length.

no free poles where magnetization is parallel to the surface. The magnetic moment of the cylinder is therefore

$$M = m \times l \quad \text{(by definition of } M\text{)}$$
$$= Jv \quad \text{(by definition of } J\text{)}$$

Therefore

$$J = \frac{ml}{v} = \frac{m}{A} \tag{8.3}$$

Thus the equivalent pole strength per unit area, m/A, is equal to the intensity of magnetization for the end surfaces of the cylinder and of course zero for the cylindrical surface. In the general case it can be shown to be equal to the component of J normal to the surface.

When calculating the anomaly in this manner the magnetic field due to each sheet of free poles is determined in turn and the total anomaly at the field point obtained by summation. The anomaly of a dipping slab (e.g. a magnetized dolerite dyke) is worked out in this way. Formulae are given in most standard texts (e.g. Telford *et al.,* 1976) and are not reproduced here, but the dyke model is a particularly important one. It is also a very useful illustration of the relationship between anomaly shape and variations in dip, strike, geomagnetic field and magnetization direction, understanding of which yields very useful insights into the behaviour of anomalies in general.

To simplify the problem and bring out important facts we consider a thin dyke infinite in depth and strike length. Variables are therefore the geomagnetic field inclination, magnetization direction of the dyke, its dip, strike and depth to top, supposed to be constant. Because the strike length is infinite the horizontal component of magnetization parallel to the dyke does not contribute to the anomaly. In calculating the anomaly, therefore, it is only necessary to consider magnetization in the plane perpendicular to the strike of the dyke. The geomagnetic field component in this plane has an inclination I'_F such that

$$\tan I'_F = \frac{\tan I_F}{\sin a} \tag{8.4}$$

where I'_F is inclination of the geomagnetic field resolved in the plane perpendicular to the dyke, I_F the true angle of inclination and a the angle the strike makes with the magnetic north. This expression combines two angular variables, inclination and strike, into one. Study of the problem then reveals that there is an interdependence of the three variables I'_P, I'_F and δ, I'_P being the inclination of the component of the resultant dyke magnetization in the vertical plane perpendicular to the strike, whilst δ is the angle of dip of the dyke (see Fig. 8.8). It is found that for any particular dyke, if we put

$$\theta = I'_F + (I'_P - \delta) \tag{8.5}$$

then provided we keep θ constant we can vary the terms on the right hand side of the expression at will without affecting the form of the anomaly calculated along a traverse across the dyke. Note particularly that for a given geomagnetic field inclination the anomaly shape is controlled by the angle between the component of magnetization I'_P and the plane of the dyke.

It is therefore possible to express the dyke anomaly as a function of θ. Thus a normalized set of anomalies can be calculated for increments of θ over its full range and this set will then cover all possible anomaly shapes for the thin dyke model, irrespective of magnetic latitude, strike, dip and direction of dyke magnetization. If therefore an observed anomaly is scaled correctly it can be compared with the calculated set and a value of θ found. The strike of the dyke a can be found from the trend of the

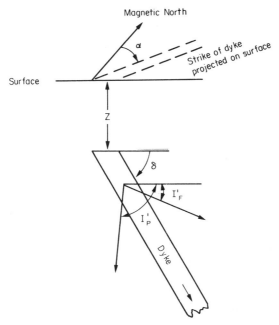

Fig. 8.8   Nomenclature for the inclined thin dyke.
   a = strike relative to magnetic north
   z = depth
   δ = dip of dyke
   I'_F = inclination of geomagnetic field in a plane perpendicular to the strike
   I'_P = inclination of the remanent magnetization in a plane perpendicular to
        the strike.

anomaly contours and knowing I_F then I'_F can be calculated. If I_P is known we can calculate I'_P and so get δ, the dip of the dyke. More often than not induced magnetization is assumed (i.e. $I_P = I_F$) and so $\theta = 2I'_F - \delta$ and the dip can still be determined. If we do not know the direction of remanent magnetization I_P cannot be calculated and the dip of the dyke is indeterminate. Three examples of thin dyke anomalies are illustrated in Fig. 8.9. A full set which includes anomalies for thick dykes is given by Parker Gay (1963).

Scaling observed anomalies both horizontally and vertically to obtain a fit to theoretical curves is not easy and various workers have calculated

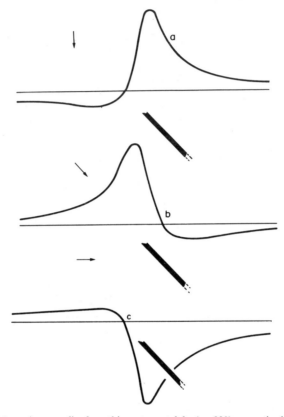

Fig. 8.9 Magnetic anomalies for a thin east–west dyke ($a = 90°$) magnetized by induction and with a dip of 45°. (a) $I_F = 90°$; (b) $I_F = 45°$; (c) $I_F = 0°$. Note that for $a = 90°$ $I_{F'} = I_F$.

characteristic curves, based on the principles discussed above, which can be used to obtain a complete interpretation without going through the process of curve matching. A very good survey of the various methods is given by Åm (1972).

8.6 Applications

Magnetic surveys have many applications in the field of mining. Ores such as magnetite and pyrrhotite, and to a less extent haematite, are strongly magnetic and thus may be directly detected with a magnetometer. To be of economic importance the ore must be of a high enough concentration, in sufficient quantity, and not too deeply buried, these being fortunately the most favourable conditions for its detection.

Many ores which are not in themselves magnetic are associated with magnetic rocks. For example, gold is sometimes found associated with certain intrusive igneous rocks which may be traced with the magnetometer. Kimberlite pipes in which diamonds are found have also been outlined by means of their magnetic anomalies.

In certain areas use has been made of the association of magnetic members of a stratigraphic series with non-magnetic beds of commercial value in order to produce a geological map from the magnetic data and thus trace the ore bearing horizons.

Though little used, magnetic surveys do have applications in the field of civil engineering. In surveying constructional sites where the solid rock is buried beneath a cover of superficial deposits surveys of this kind would be useful for tracing contacts between various types of crystalline rocks, or for tracing out on the surface the course of dykes and other intrusions which may cross the area. Such bodies may also act as impermeable barriers to water flow, so their location can be of importance in hydrogeological studies.

On a larger scale, aeromagnetic surveys are used for mapping geological structure. In areas where the sedimentary sequence is very thick it is sometimes possible to delineate the major structural features because the succession includes magnetic horizons which may be ferruginous sandstones or shales, tuffs, or possibly lava flows. In many regions, however, the igneous and metamorphic "basement" which underlies the sedimentary sequence is the predominant factor controlling the pattern of the anomaly field, for it is usually far more magnetic than the sediments. Where the basement rocks are brought nearer to the surface in structural highs the magnetic anomalies are large and characterized by strong relief. Conversely deep sedimentary basins are likely to show low values of the anomaly and gentle field gradients. In some areas of this geological type it is possible from measurements on the gradients of suitable anomalies to get a fair idea of the thickness of the sedimentary cover, but otherwise interpretation is usually qualitative. The form of the magnetic contours is often determined to a large extent by the structural trends in the basement itself and in some instances these may be of importance in that they control the later pattern of folding and faulting in the overlying sediments, though this is by no means always so. Ideally to illustrate the interpretation of aeromagnetic maps it would be necessary to discuss examples covering a very wide range of geology and structure. This is not possible here but the aeromagnetic map of the Midlands, part of which is shown in Fig. 8.10, does cover an area in which the geology is varied. It also has the advantage for our purpose of showing some anomalies which are clearly related to surface geology and some which appear to reflect major buried structure. The geomagnetic

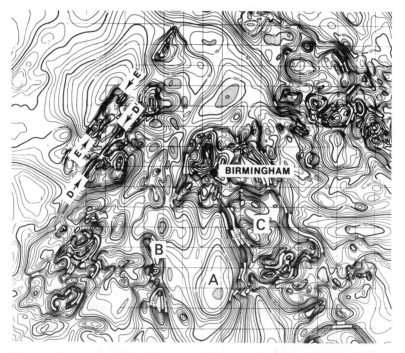

Fig. 8.10   The aeromagnetic map of part of the Midlands. Reproduced by permission of the Director, Institute of Geological Sciences; Crown Copyright reserved.

A   Worcester basin   DD′   Line of Church Stretton fault
B   Malvern horst   EE′   Line of Hodnet fault
C   Buried ridge of magnetic rocks

field is quite steeply inclined so that unless bodies have a strong remanence in a very different direction there should be a good concordance between the geology and the pattern of the magnetic anomalies, the flight line spacing of 2 km and the flight height of 300 m being sufficiently small for large scale features, as defined by areas of different magnetic "character", to be easily outlined. Areas in which magnetic rocks are at or near the surface will be characterized by marked and usually sharp anomalies. Those anomalies which are smoother or of longer "wavelength" will reflect the presence of non-magnetic rocks down to considerable depth.

The deep Worcester basin of non-magnetic Triassic sandstones and marls is dominated by smooth negative anomalies. It contrasts sharply with the strongly positive areas of sharp anomalies on either flank, indicative of near surface areas of older magnetic rocks, the steep and linear form of the gradients being in places strongly suggestive of major faulting. Magnetic rocks are seen at the surface in the Precambrian of the Malvern horst, both southerly and northerly extensions of the anomalies beyond the area of outcrop indicating the existence of these rocks at no great depth. The wide ridge of magnetic rocks trending in a roughly south-east direction from Birmingham is in general not exposed. Precambrian rocks are, however, seen in the Lickey Hills south of Birmingham and again at Nuneaton.

This interpretation is in fact confirmed by information on the deep structure that can be obtained from the Bouguer anomaly map and boreholes which leave no doubt about the existence of a major Triassic basin centred near Worcester. An east–west sketch section across the area, based on all available data, is given in Fig. 8.11, together with the aeromagnetic and gravity profiles.

The aeromagnetic map shows a wealth of other features of structural interest. Note for example the steep and linear gradient separating strongly positive and negative areas, associated with the Church Stretton fault, and how further south the fault appears to mark the westward limit of near surface magnetic rocks, probably of Precambrian age. Further to the north-west, beyond the Hodney fault, the rapidly thickening Trias of the southern parts of the Cheshire basin is revealed by a very sudden change to a smooth magnetic area characterized by a large amplitude negative anomaly.

Information of this kind is very valuable in prospecting for oil in geologically unexplored country where the first essential is to locate deep sedimentary basins favourable for its accumulation and to obtain some knowledge of their gross structure.

CHAPTER 9

Geophysical Borehole Logging

9.1 Introduction

IT is difficult to over-emphasize the value of controlling the interpretation of a geophysical survey by making use of all available geological information, and in particular of any records of boreholes or wells in the area. Such borehole records or geological logs may be less than adequate unless the hole has been cored or very careful note taken of the chippings flushed to the surface by the drilling fluid. A geophysical log of physical properties, obtained by carrying out a down-hole geophysical survey, obviously greatly increases the value of the hole to the geophysical interpreter, and is also used in a purely qualitative way to correlate beds showing a recognizable pattern of properties from one borehole to another. Such logs have been found equally valuable in estimating the reservoir properties (porosity, fluid content and approximate permeability) of oil-bearing formations and of aquifers. Geophysical logging has also been used in mining and foundation engineering, though less extensively than in reservoir evaluation. In the latter field, measurements are often also carried out on the temperature, resistivity and rate of movement of the borehole fluid: such "fluid logging" or "production logging" will not be further considered here.

All logging is carried out by lowering a detector or "sonde" on a strong multicore cable and recording its output (usually on the return journey) on a recorder at the surface with a chart driven by the cable winch. Several measurements can be made simultaneously using a multi-channel recorder, and it is now common practice to take a complete record on digital tape for subsequent processing. Sondes may be several metres in length and of a diameter suited to the common borehole diameters in the range 0.1–0.5 m: they may be free to move from side to side of the hole or held centrally in it to simplify interpretation, and in some cases the detectors are carried on spring-loaded pads which press against the borehole wall.

Logging equipment for oil wells which may be several kilometres deep is naturally on a bigger scale than that capable of measurements in water-wells down to a few hundred metres, and the more elaborate and expensive methods are still restricted to petroleum exploration. However, the importance of logging methods in groundwater evaluation is increasing, and interpretation procedures have been adapted to deal with the problems peculiar to aquifers.

In addition to the devices which measure physical properties of the formations penetrated, sondes are available to measure the parameters of the hole itself, in particular its diameter by the *caliper log* and its departure from vertical by the *well-survey* sonde. The hole diameter may increase above that of the bit ("caving") when friable formations (often, but not always, shales) are encountered, and the diameter may have to be known for quantitative log interpretation. The deviation from vertical must of course be known if the results of the survey are to be accurately plotted in space.

The readings of any logging device are influenced not only by the properties of the formation at sonde level but also, unless this formation is very thick, by the formations above and below it, and by the fluid in the borehole itself. This fluid may in some shallow holes be air, but is more often either the formation water (in some water-wells) or more commonly a *drilling mud* of thixotropic clay suspended in fresh or saline water or even in oil. The resistivity of the drilling mud may thus vary widely from one hole to another, but is commonly somewhat higher than that of the formation water in oil wells and comparable with it in water wells. Where the hole penetrates a permeable formation and the mud is dense enough for the pressure in the hole to exceed that in the formation, an *invaded zone* is formed in which the formation fluid is displaced by the *mud filtrate*. The clay in suspension in the mud is filtered off during invasion to form a *mudcake* on the hole wall. The presence of such a mudcake, which may be thick enough to be detected by the caliper log, is thus an indication of the existence of permeability, though not of its magnitude.

9.2 Electrical Logs

Electrical measurements are of obvious value in reservoir evaluation, since they can lead through a determination of the formation factor $F = \rho/\rho_w$ (see §4.2) to an estimate of the porosity $\emptyset$ through

$$F = a/\emptyset^m \tag{4.1}$$

The sonde of the various resistivity logs carries a set of electrodes recording an apparent resistivity dependent upon their geometry. Some of these electrode arrangements are shown in Fig. 9.1: the "*lateral*" and "*normal*" sondes are obviously related to the Schlumberger and Wenner configurations discussed in §4.5. The normal sonde is often recorded at both short spacing to give good resolution of thin beds and long spacing (hence deeper penetration) to give an apparent resistivity affected more

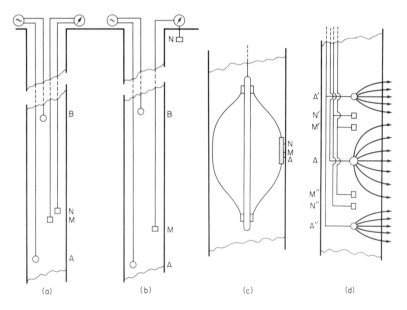

Fig. 9.1   (a) The lateral sonde, (b) The normal sonde, (c) The microlog, (d) The laterolog. A, B, current electrodes: M, N, potential electrodes. Source of low frequency alternating current. Recording potentiometer coupled to winch.

by the formations than by the borehole and invaded zone, if any. The *microlog* (Fig. 9.1(c)) on the other hand uses a spacing of only a few centimetres, giving a very good vertical resolution. Its electrodes are pressed against the hole wall by an insulating pad so that the drilling mud in the hole does not greatly contribute to the apparent resistivity recorded, which is influenced by the formation very close to the wall. Two spacings are commonly recorded together, and if the apparent resistivity is greater for the larger spacing ("positive separation") this may indicate the presence of the low resistivity mudcake and hence the existence of permeability. At the larger spacing it is the resistivity of the invaded zone ρ_i rather than that of the undisturbed formation ρ_t that is estimated from the apparent resistivity, and since this zone contains water of known resistivity ρ_{mf} (that of the mud filtrate), its formation factor can be found directly since

$$F = \rho_i/\rho_{mf} \qquad (9.1)$$

Knowing F, estimates can be made not only of the porosity of the formation but also of the resistivity ρ_w of the natural formation water, since $F = \rho/\rho_w$ and the true formation resistivity ρ_t can be found from one of the deep-penetration electrical logs.

The *laterolog* (Fig. 9.1(d)) is an example of a *focused* log in which a better compromise between penetration and resolution is achieved by the use of auxiliary current electrodes to direct the current flow into a more sheet-like form.

The *spontaneous potential* or SP is the natural potential difference existing between a single electrode in the hole and a reference electrode at the surface: it is due to a combination of electrochemical effects of which the most important is the "membrane potential" which exists across a shale-sandstone boundary. This results in a log of SP which remains relatively constant against shales, with negative excursions opposite non-shaly bands. These bands are usually permeable formations of interest as aquifers or oil-producers, which can be conveniently defined by the SP log.

The *induction log* is strictly an electromagnetic rather than an electrical log: it is the sub-surface equivalent of the Slingram system (Fig. 6.6) though the transmitting and receiving coils are arranged coaxially on the sonde. The good penetration of this system and the ease with which corrections for the borehole and invaded zone can be calculated have made it very common for estimation of true formation resistivities.

9.3 Radiometric Logs

These have the considerable advantage over electrical logs that they can be recorded after a hole has been cased. The *gamma-log* (see also Chapter 10) records natural γ-radioactivity (alpha and beta radiation have negligible penetration) within a few tens of centimetres of the borehole wall, and in a sedimentary sequence will pick out the shales, since these have higher γ-activity than other beds with the exception of potash-rich evaporites. The gamma-log is thus similar in general form to the SP log discussed above, giving an estimate of "shaliness".

The *gamma-gamma* or *density log* contains a gamma-ray source as well as a detector. The two are shielded from each other so that the detector records only the radiation scattered back from the rock formations, and the intensity of this scattering gives a good measure of the formation density. If a density can be assumed for the mineral matrix of the rock, a porosity ϕ_D can be estimated from the formation density.

The various *neutron logs* can also be used to give a porosity estimate ϕ_N for saturated formations, though the technique is a very different one. A source of high energy neutrons is held against the hole wall and the neutrons are detected at the other end of the sonde after their energy has been reduced to the equilibrium or "thermal" value by scattering within a fraction of a metre of the hole. This energy loss occurs principally by "elastic" collisions with atomic nuclei, and momentum conservation ensures that only collisions with nuclei of about the same mass as a neutron reduce its energy appreciably. In practice this means that only hydrogen nuclei contribute to the energy loss, and these are to be found almost entirely in the water or hydrocarbon present in the formation pore-space. A large hydrogen content thus leads to a rapid neutron energy loss, and so to a low reading at the detector, from which the porosity estimate ϕ_N follows directly without the intermediate step of finding the formation factor.

Different versions of the sonde detect the neutrons of near thermal energies in different ways: either directly (the *neutron-neutron* log) or as the gamma radiation emitted when thermal neutrons are finally captured by nuclei (the *neutron-gamma* log). The *neutron lifetime log* is a further development which uses a pulsed neutron source and estimates an average "capture cross-section" for the formation from the time taken (of the order of a millisecond) for the neutron cloud to decay. This capture cross-section is large for hydrogen (hence for water and petroleum) and chlorine (hence for saline water) and small for low-porosity sandstones and limestones. The log readings therefore tend to correlate well with those of resistivity logs.

9.4 Sonic Logging

The sonic or continuous velocity log (CVL) measures the transit time of an ultrasonic pulse through a metre or two of formation, and thus gives seismic velocity as a function of depth with a much finer vertical resolution than can be achieved from a refraction or even from a reflection seismic survey. The sonde normally carries a transmitter at each end, and these are pulsed alternately, the pulses being received by a spaced pair of transducers between them. The four travel times can be used to form "minus" times (§3.52) to give a velocity unaffected by the variable delay times in the borehole. Apart from its value in providing a control velocity-depth function for seismic interpretation, the sonic log can be used to make a third estimate of porosity ϕ_S by using the time-average equation

$$\frac{1}{V} = \frac{\phi}{V_f} + \frac{1 - \phi}{V_m} \tag{2.18}$$

(see §2.4) with an assumed value for the matrix velocity V_m. Different lithologies will obviously have true V_m values which differ systematically from that assumed, and so ϕ_S will depart from the true porosity in a way which systematically depends on lithology. The estimates ϕ_D and ϕ_N are also systematically in error for comparable reasons, and it is useful to form a *cross-plot*, such as that shown in Fig. 9.2, of two of the measurements from which porosity estimates are made, or of the estimates themselves. The observations are found to fall not on a single line but on a family of lines each of which corresponds to a particular lithology, so

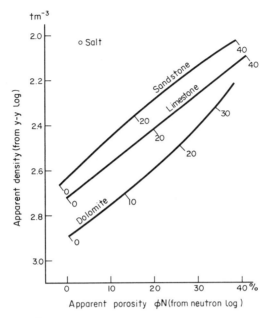

Fig. 9.2 An example of a "cross-plot" of log readings, using the density estimated from the γ-γ log and the porosity ϕ_N from the neutron log. Each line is calibrated empirically in terms of true porosity. Note the anomalous position of salt, which is thus readily identifiable.

that from a calibrated cross-plot both lithology and true porosity can be estimated from a pair of log readings.

9.5 Temperature Logging

If the temperature θ_z is measured at a depth z in a borehole, or better the temperature gradient $d\theta_z/dz$ is measured directly by a closely spaced pair of thermistor detectors, then in principle the thermal conductivity k_z can be found as a function of depth since

$$d\theta_z/dz = H/k_z \qquad (9.2)$$

where H is the upward heat flow, supposed uniform, from the interior of the earth. In practice however the situation is much more complicated: in the first place, temperature gradients near the surface are strongly affected by the daily and annual temperature cycles imposed by solar heating, which become negligible only at a depth of 20 m or so. Secondly, the heat flow H may not be uniform with depth, since in regions of groundwater circulation it is modified by the heat carried convectively by the water. Thus at best only approximate and relative conductivity estimates are possible, but these may sometimes be of value in picking out beds of anomalous conductivity such as coal seams. Thermal measurements in boreholes are, however, more commonly combined with measured conductivity values from core samples to give estimates of heat flow which are of interest in surveys to assess the practicability of geothermal power utilization.

CHAPTER 10

Radiometric Surveys
and Remote Sensing

10.1 Radiometric Surveys

As well as differing in their densities and magnetic properties, rocks vary widely in their content of radioactive elements, and this variation is the basis of a third "passive" method of geophysical prospecting in which natural radioactivity is detected in either ground-based or airborne surveys.

Although many elements are naturally radioactive, only three contribute appreciably to the radiation field that is measured. They are uranium, thorium and potassium, or rather the active isotope of potassium K^{40} which is present to the extent of about 0.01 per cent of naturally occurring potassium. The radioactive decay of K^{40} is relatively simple, resulting either in a transition to a calcium isotope of the same mass by emission of an electron (or "β-particle") or in the formation of argon A^{40} by addition of an electron and simultaneous emission of a gamma-ray photon of fixed energy 1.46 MeV. These two decay processes occur in a fixed ratio known as the branching ratio.

The radioactivity of uranium and thorium is more complex, since each of these elements decays to another which is itself unstable and this process is repeated to give three radioactive decay series (there are two naturally occurring isotopes of uranium) each ending in one of the stable isotopes of lead. The radiation emitted in the course of each series includes alpha-particles (high-energy helium nuclei) as well as β-particles and γ-radiation of various energies.

The charged particles constituting a and β radiation lose energy rapidly in passing through matter by ionizing it, and consequently have a range of the order of millimetres or less in solids. Even in air, the range of β radiation is less than a metre, and that of α radiation much less, so that neither is of real value for prospecting. Gamma-radiation, on the other hand, is simply electromagnetic radiation of extremely short wavelength, loses much less energy by ionization, and therefore has a much greater range of the order of 100 m in air, so that variations in surface radiation can be detected from low-flying aircraft. It is, however, *surface γ-radiation* that is measured, since even this radiation can penetrate no more than a metre of rock.

Gamma-radiation is normally detected with a *scintillometer,* consisting of a crystal (sodium iodide "activated" with thallium) which responds to the absorption of a γ-ray photon by the emission of a flash of visible light. This scintillation is converted to an electrical pulse by a photo-multiplier device, and the amplitude of the pulse is approximately proportional to the energy of the γ-photon. The rate at which pulses are recorded depends on the intensity of the γ-ray flux and on the size of the crystal intercepting it. The intensity in turn depends on the solid angle which the source subtends at the detector, so that to detect localized or weak sources of radioactivity from the air requires a large scintillometer crystal, several tens of centimetres in diameter. The pulses are averaged over a short time sufficient to smooth out statistical fluctuations, and the count-rate is recorded on a strip-chart and often also numerically, as in an aeromagnetic survey, which is usually conducted at the same time as a radiometric survey to aid in the interpretation of both. Aircraft position is of course monitored by photography as with all airborne surveys.

Interesting radiometric anomalies recorded from the air may be followed up on the ground with portable instruments using smaller crystals. Both portable and airborne scintillometers are now commonly coupled to *gamma-ray spectrometers* which sort the electrical pulses from the photomultiplier according to their amplitude into three or more ranges before recording separately the rate of arrival of pulses of each amplitude range. Since pulse amplitude depends upon gamma-ray energy, the various spectrometer channels are associated with different energy bands, and can be so chosen that the counting rates recorded are mainly associated with uranium, thorium and potassium radiation respectively, though the complexity of the uranium and thorium gamma-ray spectra makes a complete separation impossible.

Even with the additional information provided by a spectrometer, the interpretation of radiometric surveys remains at best semi-quantitative. Highs and lows of measured γ-radiation correspond to areas of more or less active rock very near to the ground surface. The only way in which active material buried by more than a metre or two can contribute to surface activity is by a process of migration to the surface, for example of the gas radon which forms as an intermediate product in the decay of uranium and thorium. A low in recorded activity may thus be due to an increase in overburden thickness rather than to a change in the bedrock itself, and similarly a high may indicate an area such as a fault zone where upward migration of radon is facilitated.

Particularly high radioactivity is of course shown by those ores of uranium and thorium which are one of the chief targets of radiometric surveys. However, the method is also used as an aid to geological mapping, particularly in basement areas where granitic rocks can often be distinguished from others containing less potassium felspar. Sedimentary rocks, except potash-rich evaporites, are of lower activity, but boundaries between sandstones and more strongly active shales may be defined.

10.2 Remote Sensing

The airborne survey techniques which have come to be grouped under this name are similar to the radiometric method in that they gather information only from the ground surface and have none of the subsurface penetration which is the essence of other geophysical methods. A radiometric survey, however, is carried out along discrete profiles which are then interpreted separately or compiled into a contour map: the methods discussed below are more akin to aerial photography in that they provide a high-resolution record of the surface in two dimensions, either by some variation of conventional aerial photography or by a scanning process which builds up continuous coverage of the ground during a single passage of the surveying aircraft in the same way as a television picture.

As one would expect, aerial photographs in colour can be more readily interpreted in terms of ground features than those in black and white, and another field of interpretation is opened when infra-red radiation from the ground is recorded in addition to the whole of the visible spectrum. Regions which appear bright in this part of the spectrum are either at a higher temperature than their surroundings, which may in itself be of interest, or of higher infra-red reflectance, a property which is sensitive to the nature of vegetation cover. The cover of natural vegetation is often in its turn strongly influenced by geology, so that photogeological interpretation may be aided by infra-red sensing.

Surveys of this kind, which include those continuously carried out by the Landsat satellites, often record simultaneously in several spectral bands either by photography or by electronic recording through a set of optical filters. The intensities recorded in the various filtered bands can then be displayed as a "false-colour" image in which, for example, the infra-red band may be represented as red, the red as green, and so on. Electronic recording in particular obviously lends itself to variations in the final display designed to accentuate features of particular interest. One common variation is to select specific amplitude ranges or "density slices" of the response in one wavelength band and to display them in coded form with each density slice appearing as a patch of a different colour on the final presentation, which can thus be made to correspond more closely with, for example, divisions between types of vegetation.

Applications of this broad-band aerial photography are most numerous in fields such as studies of land utilization, but photogeological mapping has also been aided either by association of geological boundaries with those between different forms of natural vegetation, or more directly in areas of exposed rock by making use of differences between rock types in their thermal response to daily temperature variations.

A very different development in airborne mapping is the use of radar to "illuminate" the ground so that a survey can be carried out in cloudy conditions or at night when conventional photography is impossible. The technique is known as Side Looking Airborne Radar or SLAR, and is analogous to the marine side-scan sonar system mentioned in §3.2. A narrow radar beam is emitted at right angles to the track of the aircraft, and echoes are received from the ground vertically below and up to 100 km away along the narrow strip illuminated by the beam. As the aircraft moves, returns from successive strips are built up into a complete two-dimensional coverage with a resolution of the order of 5 m. The oblique illumination of the ground gives the enhancement of relief seen in aerial photographs taken near dawn or dusk, and any conductors such as man-made metallic objects act as particularly strong radar reflectors and are therefore clearly shown. The advantage of the method lies in its ability to cover large areas rapidly and cheaply, but it is not capable of the resolution shown in the best conventional surveys.

Bibliography

THIS list includes papers specifically referred to in the text and other key references for each chapter, but is far from comprehensive. The books given in the "general" section below are good starting points for a more detailed literature search.

GENERAL

CLARK, S. P. (Ed.) (1966). *Handbook of Physical Constants,* Geol. Soc. Am. Memoir 97 (a compilation of physical properties of rocks).

DOBRIN, M. B. (1976). *Introduction to Geophysical Prospecting,* 3rd edn., McGraw-Hill.

GRANT, F. S. and G. F. WEST (1965). *Interpretation Theory in Applied Geophysics,* McGraw-Hill.

Mining Geophysics I & II (1967) (Society of Exploration Geophysicists).

MORLEY, L. W. (Ed.) (1970). *Mining and Groundwater Geophysics* (Economic Geology Rep. 26, Geological Survey of Canada).

PARASNIS, D. S. (1979). *Principles of Applied Geophysics,* 3rd edn., Chapman & Hall.

SHERIFF, R. E. (1973). *Encyclopaedic Dictionary of Exploration Geophysics* (Society of Exploration Geophysicists).

TELFORD, W. M., L. P. GELDART, R. E. SHERIFF and D. A. KEYS (1976). *Applied Geophysics,* Cambridge University Press.

Geoexploration (quarterly), Elsevier.

Geophysical Prospecting (quarterly, European Association of Exploration Geophysicists).

Geophysics (monthly, Society of Exploration Geophysicists).

CHAPTER 2

BIRCH, F. (1961). The velocity of compressional waves in rocks to 10 kilobars, *J. Geophys. Res., 66,* 2199–224.

HOLTZSCHERER, J. J. and G. DE Q. ROBIN (1954). Depth of polar ice caps. *Geog. J., 120,* 193–202.

TALWANI, M., J. L. WORZEL and M. EWING (1961). Gravity anomalies and crustal structure across the Tonga Trench, *J. Geophys. Res., 66,* 1265–78.

WHITE, J. E. (1965). *Seismic Waves—Radiation, Transmission and Attenuation,* McGraw-Hill.

CHAPTER 3

ANSTEY, N. A. (1970). Seismic prospecting instruments, Vol. 1: Signal characteristics and instrument specification. *Geoexploration Monographs,* Ser. 1, No. 3, Gebrüder Borntraeger.

CRAWFORD, J. M., W. E. N. DOTY and M. R. LEE (1960). Continuous signal seismograph, *Geophysics,* **25**, 95–105.

EVENDEN, B. S. and D. R. STONE (1971). Seismic Prospecting Instruments, Vol. 2: Instrument performance and testing, *Geoexploration Monographs,* Ser. 1, No. 3, Gebrüder Borntraeger.

FITCH, A. A. (1976). Seismic reflection interpretation. *Geoexploration Monographs,* Ser. 1, No. 8, Gebrüder Borntraeger.

FRENCH, W. S. (1974). 2D and 3D migration of model experiment reflection profiles, *Geophysics,* **39**, 265–77.

GREEN, R. (1962). The hidden layer problem, *Geophys. Prosp.,* **10**, 166–70.

HAGEDOORN, J. G. (1959). The Plus–Minus method of interpreting seismic refraction sections, *Geophys. Prosp.,* **7**, 158–83.

MCQUILLIN, R., M. BACON and W. BARCLAY (1979). *An Introduction to Seismic Interpretation,* Graham & Trotman.

MARR, J. D. (1971). Seismic stratigraphic exploration, *Geophysics,* **36**, 311–29, 533–53 and 676–89.

MAYNE, W. H. (1967). Practical considerations in the use of common reflection point techniques, *Geophysics,* **32**, 225–9.

MUSGRAVE, A. W. (Ed.) (1967). *Seismic Refraction Prospecting* (Society of Exploration Geophysicists).

NEITZEL, E. B. (1958). Seismic reflection records obtained by dropping a weight, *Geophysics,* **23**, 58–80.

PAYTON, C. E. (Ed.) (1977). *Seismic Stratigraphy: Applications to Hydrocarbon Exploration* (Am. Assoc. of Petr. Geol. Mem. 26).

SCHNEIDER, W. A. (1971). Developments in seismic data processing and analysis (1968–70), *Geophysics,* **36**, 1043–73.

SOSKE, K. L. (1959). The blind zone problem in engineering geophysics, *Geophysics,* **24**, 359–65.

TROREY, A. W. (1970). A simple theory for seismic diffractions, *Geophysics,* **35**, 762–84.

TUCKER, P. M. and H. J. YORSTON (1973). *Pitfalls in Seismic Interpretation* (Society of Exploration Geophysicists).

CHAPTER 4

BHATTACHARYYA, P. K. and H. P. PATRA (1968). *Direct Current Geoelectric Sounding. Principles and Interpretations,* Elsevier.

COMPAGNIE GÉNÉRALE DE GÉOPHYSIQUE (1955). Abaques de sondage électrique, *Geophys. Prosp.,* Supp. No. 3.

GHOSH, D. P. (1971). Inverse filter coefficients for the computation of apparent resistivity standard curves for a horizontally stratified earth, *Geophys. Prosp.,* **19**, 769–75.

HABBERJAM, G. M. and G. E. WATKINS (1967). The use of a square configuration in resistivity prospecting, *Geophys. Prosp.,* **15**, 445–67.

KOEFOED, O. (1960). A generalized Cagniard graph for the interpretation of geoelectrical soundings, *Geophys. Prosp.,* **8**, 459–73.

KOEFOED, O. (1968). The application of the kernel function in interpreting geoelectrical resistivity measurements. *Geoexploration Monographs,* Ser. 1, No. 2, Gebrüder Borntraeger.

KUNETZ, G. (1966). Principles of direct current resistivity prospecting, *Geoexploration Monographs,* Ser. 1, No. 1, Gebrüder Borntraeger.

MARSDEN, D. (1973). The automatic fitting of a resistivity sounding by a geometrical progression of depths. *Geophys. Prosp.* **21,** 266–80.

MOONEY, H. W. and W. W. WETZEL (1956). *Potentials about a Point Electrode and Apparent Resistivity Curves for a Two-, Three- and Four-layer Earth.* University of Minnesota Press.

ORELLANA, E. and H. M. MOONEY (1972). *Two and Three Layer Master Curves and Auxiliary Point Diagrams for Vertical Electrical Sounding using Wenner Arrangement.* Interciencia.

ROY, A. and A. APPARAO (1971). Depth of investigation in direct current methods. *Geophysics,* **36,** 943–59.

ZOHDY, A. A. R. (1975). *Automatic Interpretation of Schlumberger Sounding Curves using Modified Dar Zarrouk Functions,* US Geol. Survey Bulletin 1313E, Washington.

CHAPTER 5

BARKER, R. D. (1978). The reduction of lateral effects in resistivity sounding, *Geophys. J.R. astr. Soc.,* **53,** 143–4.

CARPENTER, E. W. and G. M. HABBERJAM (1956). A tri-potential method of resistivity prospecting, *Geophysics,* **21,** 455–69.

FLATHE, H. (1970). Interpretation of geoelectrical resistivity measurements for solving hydrogeological problems, *in* L. W. MORLEY (Ed.). *Mining and Groundwater Geophysics* (Economic Geology Rep. 26, Geological Survey of Canada).

HABBERJAM, G. M. (1979). Apparent resistivity observations and the use of square array techniques, *Geoexploration Monographs,* Ser. 1, No. 9, Gebrüder Borntraeger.

HABBERJAM, G. M. and A. A. JACKSON (1974). Approximate rules for the composition of apparent resistivity sections, *Geophys. Prosp.,* **22,** 393–420.

HATHERTON, T., W. J. P. MACDONALD and G. E. K. THOMPSON (1966). Geophysical methods in geothermal prospecting in New Zealand, *Bull. Volcan.,* **29,** 485–97.

VAN DAM, J. C. and J. J. MEULENKAMP (1967). Some results of the geoelectrical resistivity method in groundwater investigations in the Netherlands, *Geophys. Prosp.,* **15,** 92–115.

VAN NOSTRAND, R. G. and K. L. COOK (1967). *Interpretation of Resistivity Data.* US Geol. Survey Prof. Paper 499.

CHAPTER 6

BERTIN, J. and J. LOEB (1976). Experimental and theoretical aspects of Induced Polarization, *Geoexploration Monograph,* Ser. 1, No. 7, Gebrüder Borntraeger.

COGGAN, J. H. (1971). Electromagnetic and electrical modelling by the finite element technique, *Geophysics,* **36,** 137–55.

HOOD, P. J. and S. H. WARD (1969). Airborne geophysical methods, *Adv. in Geophys.,* **19,** 1–41.

KELLER, G. V. and F. C. FRISCHKNECHT (1966). *Electrical Methods in Geophysical Prospecting,* Pergamon Press.

PARASNIS, D. S. (1973). *Mining Geophysics,* Elsevier.

PEMBERTON, R. H. (1962). Airborne E.M. in review. *Geophysics,* **27,** 691–713.

SUMNER, J. S. (1976). *Principles of Induced Polarization for Geophysical Exploration,* Elsevier.

WARD, S. H. (1967). In *Mining Geophysics,* Vol. 2, pp. 224–372 (Society of Exploration Geophysicists).

WARD, S. H. (1970). Airborne electromagnetic methods, *in* L. W. MORLEY (Ed.). *Mining and Groundwater Geophysics* (Economic Geology Rep. 26, Geological Survey of Canada).

CHAPTER 7

BLUNDELL, D. J., D. H. GRIFFITHS and R. F. KING (1969). Geophysical investigations of buried river valleys around Cardigan Bay, *Geol. J.,* **6,** 161–80.

BOTT, M. H. P. (1959). The use of electronic digital computers for the evaluation of gravimetric terrain corrections, *Geophys. Prosp.,* **7,** 45–54.

BOTT, M. H. P. and D. MASSON-SMITH (1960). A gravity survey of the Criffel granodiorite and the New Red Sandstone deposits near Dumfries, *Proc. Yorks. geol. Soc.,* **32,** 317–32.

GARLAND, G. D. (1965). *The Earth's Shape and Gravity,* Pergamon Press.

HAMMER, S. (1939). Terrain corrections for gravimeter stations, *Geophysics,* **4,** 184–94.

HEISKANEN, W. A. and F. A. VENING MEINESZ (1958). *The Earth and its Gravity Field,* McGraw-Hill.

MESKÓ, A. (1965). Some notes concerning the frequency analysis for gravity interpretation, *Geophys. Prosp.,* **13,** 475–88.

NETTLETON, L. L. (1976). *Gravity and Magnetics in Oil Prospecting,* McGraw-Hill.

OKABE, M. (1979). Analytical expressions for gravity anomalies due to homogeneous polyhedral bodies and translations into magnetic anomalies, *Geophysics,* **44,** 730–41.

TALWANI, M. (1973). Computer usage in the computation of gravity anomalies, *Methods in Computational Physics,* **13,** 343–64.

TALWANI, M., J. L. WORZEL and M. LANDISMAN (1959). Rapid gravity computation for two-dimensional bodies with applications to the Mendocino submarine fracture zone, *J. Geophys. Res.,* **64,** 49–59.

CHAPTER 8

ÅM, K. (1972). The arbitrarily magnetized dyke; interpretation of characteristics, *Geoexp.,* **10,** 63–90.

IAGA (1976). International Geomagnetic Reference Field, 1975.0. EOS, **57,** 120–1.

PARKER GAY, S. Jnr. (1963). Standard curves for the interpretation of magnetic anomalies over long tabular bodies, *Geophysics,* **28,** 161–200.

REFORD, M. S. and J. S. SUMNER (1964). Aeromagnetics, *Geophysics,* **29,** 482–516.

SMELLIE, D. W. (1956). Elementary approximations in aeromagnetic prospecting, *Geophysics,* **21,** 1021–40.

TARLING, D. H. (1971). *Principles and Applications of Palaeomagnetism,* Chapman & Hall.

VACQUIER, V., N. C. STEENLAND, R. G. HENDERSON and I. ZEITZ (1951). Interpretation of Aeromagnetic Maps, *Geol. Soc. Am.,* Mem. 47.

CHAPTER 9

Various Authors (1970). Borehole geophysics symposium, *Geophysics,* **35,** 81–152.
WYLLIE, M. R. J. (1963). *Fundamentals of Well Log Interpretation,* Academic Press.

CHAPTER 10

SWAIN, P. H. and S. M. DAVIS (Eds.) (1978). *Remote Sensing, The Quantitative Approach,*
McGraw-Hill.

Problems

THE only way to fully appreciate the finer points of the geophysical methods described in this book is to apply them to real data and check the interpretation by drilling. Examples using real data are usually too open-ended to present satisfactorily in an introductory text, and we have preferred to include a selection of synthetic problems of the kind that we have found useful in teaching the subject. They do not by any means cover all the topics, nor even all the chapters, of the book, but should give some feeling for the attitude of mind that has to be adopted in geophysical interpretation.

Questions

Q.1. In an area of unknown geology an unreversed seismic refraction line AG has given the first arrival travel-time plot sketched in Fig. Q.1(a) (see page 210).

(a) Which four of the schematic true-scale geological cross-sections (Figs. Q.1(c)–(h)) are approximately consistent with this plot and which are totally inconsistent with it? Give reasons for rejecting the sections which you think are inconsistent with the data and velocity values for those you accept.

(b) A subsequent high-resolution reflection profile leads to the seismic reflection section of Fig. Q.1(b). Which of the possible geological sections is now indicated to be the correct one? Explain all the events on the reflection section, and suggest reasons for the amplitude variation indicated qualitatively by line thickness.

(c) Sketch the reversed $t - x$ plot $G \rightarrow A$ that you would expect for the geological section now taken to be correct.

Q.2. A seismic reflection section shows a set of parallel horizontal reflectors at two-way times of 1.00, 2.00, 3.00, and 4.00 s. The time-average velocities for these reflection times have been estimated from stacking velocities at each end of the section to be 2.50, 3.00, 3.50, and 4.00 km s^{-1} respectively.

(a) Find the depth to each reflector and note the vertical compression of the seismic section due to the increase of velocity with depth.

(b) Find the depth separations of pairs of reflectors 100 ms apart centred at 1, 2, 3, and 4 s.

(c) If the scale of the seismic section is such that 1 cm vertically represents 100 ms two-way time and 1 cm horizontally represents 100 m on the ground, find the vertical exaggeration of the section at 1, 2, 3, and 4 s.

(d) If the average velocity to 2.00 s in the central part of the section is found to be only 2.90 km s^{-1}, how much relief is actually present in this apparently "flat" reflector?

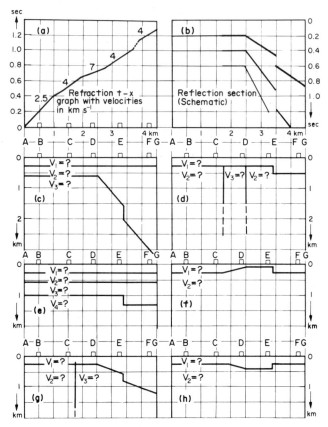

Fig. Q.1

Q.3. Along the line covered by the seismic reflection section shown in Fig. Q.3, Tertiary shales are known to outcrop at A and D and to be underlain by horizontal Carboniferous sandstones, which in turn lie conformably on limestones. As the equipment could not record reflections with two-way times less than 150 ms, reversed refraction lines were shot along AB and CD to provide information about the shallow structure. Both lines gave symmetrical $t-x$ graphs with branches corresponding to $V_1 = 2$ m ms^{-1} and $V_2 = 3$ m ms^{-1}, with intercept times at A and B of 75 ms and at C and D of 38 ms.

Interpret the refraction lines and use the results to convert the reflection time section into a depth section, taking the sandstones to be of uniform velocity.

Draw the $t-x$ graphs that you would expect from a reversed refraction line along DE.

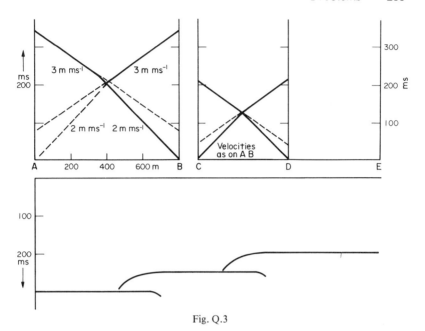

Fig. Q.3

Q.4. A "sparker" seismic reflection profiling survey has been conducted along a line from the open sea towards the coast beyond B (Fig. Q.4), crossing a sandbank which it is suspected may conceal a former cliff line. Interpret the reflection events 1, 2, and 3 as quantitatively as possible, making use of the fact that a refraction profile from A to B gave first-arrival $t-x$ branches with velocities 1.5, 2, and 3.5 m ms^{-1}. Construct the refraction $t-x$ graph that would be consistent with your reflection interpretation.

Q.5. (*a*) A reversed E–W seismic refraction line 100 m long across the western edge of a drift-filled valley running N–S gave results shown in Fig. Q.5. Interpret these travel-time graphs.

A vertical borehole at B meets granite at 5 m: does this confirm your interpretation?

(*b*) Another shot is now fired from A into a 400 m spread covering the full width of the valley. Travel times are as follows:

Range east of A (m)	50	100	150	200	250	300	350	400
Travel time (ms)	10	23	36	51	64	75	75	80

Find delay times at the geophones from the above data, using equation (3.7) and the known delay time at A.

Convert these delay times to depths and so construct the cross-section of the valley, stating any assumptions that you have to make.

Q.6. Apparent resistivity is measured with a constant electrode separation on a traverse E–W across a N–S trending vertical fault between low resistivity shales (ρ_2) and high

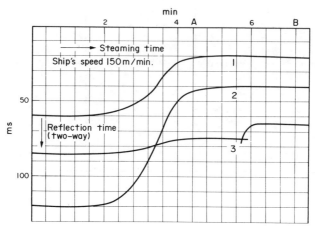

Fig. Q.4

resistivity limestones. As shown in Fig. Q.6, a three-electrode configuration is used in which the separation of the potential electrodes is small relative to their distance from the current electrode and the current electrode leads the potential electrodes. The second current electrode is at a distance, effectively at infinity. For the array positions shown in the figure draw sketches showing the positions of any images and give their strengths and polarities. Using these as a guide, plot the variation of the apparent resistivity as the array is traversed across the fault. Be as accurate as possible, without calculating the data. In assessing the effect of a boundary it will be helpful to remember that the contribution of

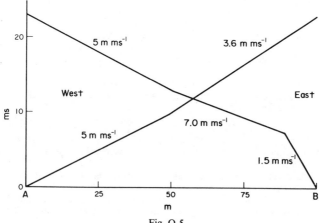

Fig. Q.5

West ◄——— East

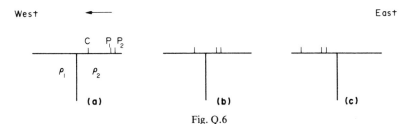

Fig. Q.6

an image to the measured potential gradient will make this anomalously high or low depending on whether it adds to or subtracts from that due to the current source. As a consequence, the calculated apparent resistivity will be higher or lower than the true value for the medium.

Q.7. Sketch as accurately as possible the apparent resistivity curve for a Wenner sounding made over ground consisting of three layers with $\rho_1 = 100$ ohm-m, $h_1 = 10$ m; $\rho_2 = 33$ ohm-m, $h_2 = 20$ m and $\rho_3 = 300$ ohm-m. Draw firstly a two-layer curve, assuming the second layer is infinitely thick. Then calculate the replacement resistivity and thickness (see p. 95) for the top two layers and draw a second two-layer curve with the replacement layer as the top layer. Finally, sketch in the complete three-layer curve. The two-layer curves of Fig. 4.9 may be used as a guide.

Q.8. Apparent resistivity – depth soundings were made at A, B, C, and D along a traverse as shown in Fig. Q.8.1. Also plotted in the figure is the Bouguer anomaly obtained from gravity measurements along the line. The apparent resistivity curves are shown in Fig. Q.8.2. A borehole not far from A passed through dry sands and reached shales at 12 m. Some distance beyond D limestones outcrop. Give a qualitative interpretation of the data, and sketch the form of a Wenner constant separation traverse along the traverse made with the spread perpendicular to the line. Say what spacing you would choose to obtain the most information about the position of the change in bedrock lithology.

Q.9. A resistivity survey has been carried out over a thick and uniform sandstone formation, the objective being to map a saline water layer in the rock at depth. The average resistivity of the saline layer was found to be 77 ohm-m. Laboratory measurements on the sandstone over a range of salinities indicated a constant formation

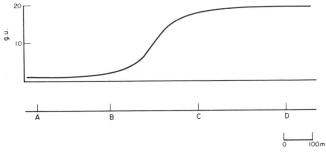

Fig. Q.8.1

Resistivity soundings at A,B,C and D

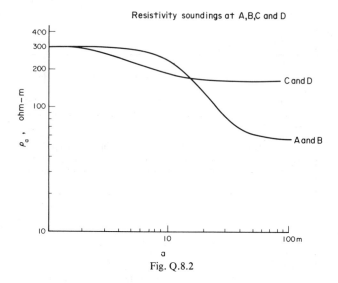

Fig. Q.8.2

factor. The typical saturated density for the sandstone was found to be around 2.34 t m⁻³ Assuming the matrix density to be 2.67, calculate the porosity ϕ. Taking $F = \phi^{-1.5}$ as a good approximate to the relationship between formation factor and fractional porosity and using the conductivity data given in the graph of Fig. Q.9, calculate the salinity of the water in the sandstone.

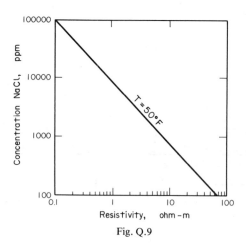

Fig. Q.9

Q.10. A gravity survey in a flat onshore area has delineated an approximately linear anomaly, the cross-profile of which is given in Fig. Q.10.

The geology can be broadly subdivided into a Tertiary sequence of clays, of average density 2.2 t m^{-3}, separated by a major unconformity from a thick sequence of Triassic rocks. The latter are dominantly sandstones, usually flat-lying, for which a density value of 2.4 t m^{-3} has been obtained, though halokinetic features are not unknown.

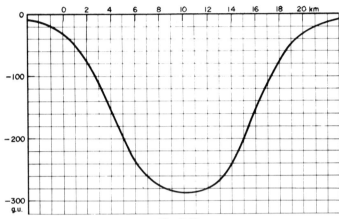

Fig. Q.10

On the geological information, the gravity low was thought to indicate a Tertiary basin. Estimate the maximum depth of the basin, assuming that the gravity anomaly arises from the sandstone/clay contrast across the unconformity.

A subsequent seismic reflection line across the basin was interpreted to give depths to the unconformity as follows:

Distance (km)	0	2	4	6	8	9	10	11	12	14	16	18	20
Depth (m)	200	620	1340	1910	1490	1000	800	1000	1490	1910	1340	620	200

Calculate approximately the gravity anomaly which would arise from this relief on the unconformity and revise your interpretation of the observed anomaly in the light of the seismic results. Does your revised interpretation enhance or detract from the value of the basin as a hydrocarbon prospect?

Q.11. To investigate the nature of a faulted boundary between Permo-Triassic rocks and unexposed, but presumed, Coal Measures, between which the density contrast σ is expected to be -0.25 t m^{-3}, a gravity survey has been carried out. The Bouguer anomaly resulting from one of a series of profiles which were run perpendicular to the supposed trend of the faulting, and which all gave essentially similar results, is given in Fig. Q.11. To the east and west respectively the anomaly tends to values of 20.0 g.u. and 5.0 g.u. Find the simplest interpretation of the data, assuming a single vertical fault, and check whether the distances to the $g_{max}/4$ and $3g_{max}/4$ points (Fig. 7.8) are consistent with it.

Subsequent drilling to the east of *B* indicates that, to the east of *C*, the Coal Measures are covered by 25 m of Permo-Trias. At *C* a vertical fault downthrows the unconformity 50 m to the west. The drilling also shows the unconformity to be horizontal and the density contrast between the now-proved Coal Measures and the Permo-Trias to be 0.238 t m^{-3}. In the light of this additional information, revise your interpretation by calculating the anomaly for the proved fault and subtracting it from that observed. (The anomaly for a thin, horizontal semi-infinite sheet may be written, using eqn. (7.12):

x/z	-4	-3	-2	-1	0	1	2	3	4
g/g_{max}	.08	.10	.15	.25	.5	.75	.85	.90	.92

where z is the depth to the sheet, x is the horizontal distance measured from vertically above the sheet edge, and positive over the sheet, and $g_{max} = 2\pi G\sigma t = 0.42\sigma t$, where σt is measured in t m^{-2} and g_{max} in g.u.)

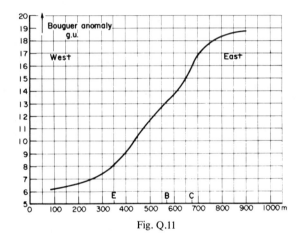

Fig. Q.11

A further drill-hole at *E* was abandoned at 60 m, still in Permo-Trias. It had, however, penetrated an extensive thickness (20 m in total) of salt horizons. Would this evidence cause you to modify your previous interpretation, and, if so, in what direction?

Q.12. Using the approach of Fig. 8.6, sketch vertical component and horizontal component magnetic anomalies for:

(a) a near-spherical body with a strong remanence directed vertically downwards,

(b) a volcanic neck, magnetized *upwards*,

(c) a horizontal sheet-like body magnetized vertically downwards, and terminated by a fault.

In all three cases take the geomagnetic field to be inclined steeply downwards.

Answers

A.1. (*a*) Of the possible geological sections, (c) cannot be accepted since head wave ray paths will reach the surface within the line *AG* only from the uppermost part of the 45° slope, and certainly not from the vertical step which would otherwise account for the increase in delay time between *E* and *F*. Section (d) is acceptable with the reservation (applying also to (c), (e) and (g)) that 7 km s^{-1} is a high velocity to find at such a shallow depth. Section (e) with the low-velocity V_4 underlying V_3 is clearly unacceptable, but (f) seems to be very plausible, explaining the anomalously high V_3 as an up-dip apparent velocity from the V_2 refractor. Section (g) is as acceptable as (d), simply explaining the second 4 km s^{-1} branch as a down-dip version of V_3, but (h) cannot be correct, since the third $t-x$ branch would have to show a down-dip apparent velocity from V_2 of less than 4 km s^{-1}. We are thus left with (d), (f) and (g), the preferred solution being (f).

(*b*) However, the reflection section (Fig. Q.1(b)) gives us a choice between (c) and (g) only. The former, even if taken to be consistent with the refraction data, should give a flat-lying reflector from *A* to *G* at a two-way time of $2 \times 250/2.5 = 200$ ms, and the $V_2 - V_3$ horizon with its dip from *D* to *G* should give a reflector (from *A* to *D*) at a two-way time of $200 + (2 \times 350/4) = 375$ ms. Neither feature is seen in the reflection section, but section (g) should give a reflector of the form of the uppermost one in Fig. Q.1(b), the later and weaker reflections being explicable as multiples. The increase in reflection amplitude between *C* and *D* can be accounted for by the likely increase in reflection coefficient (which cannot be evaluated without knowledge of the densities) from the $V_1 - V_3$ reflection.

(*c*) The reversed $t-x$ plot from *G* to *A* will start with a branch of velocity $V_1 = 2.5$ km s^{-1}, which might be expected to be followed by one showing a high apparent velocity V_u corresponding to the down-dip velocity $V_d = 4$ km s^{-1} shown on the forward profile for the dipping part of the 7 km s^{-1} refractor. (Using equations (2.16) and (2.17) it can easily be shown that V_u would be about 41 km s^{-1}.) However, the cross-over range between the V_1 and V_u branches must be at least 2 km (though not much more because the high value of V_u means that the factor $\sqrt{(V_u + V_1)/(V_u - V_1)}$ which must be used in equation (2.10) is nearly unity) and therefore no part of the dipping refractor will be "seen" as a first arrival. The second branch of the $t-x$ graph will show the true velocity V_3, changing to $V_2 = 4$ km s^{-1} near *C*.

A.2. (*a*) Depths given by $\bar{V}t/2$ are (using $\bar{V} = 2500$ ms^{-1}, etc.) 1250, 3000, 5250, 8000 m.

(*b*) Taking $\bar{V}$ to be constant over the 100 ms reflection time interval, the depth separation will be 50 ms $\times \bar{V}$ m ms^{-1}, i.e. 125, 150, 175, and 200 m, again showing the scale change of the seismic section.

(*c*) Using the result of (*b*) it is easily seen that the vertical exaggeration is 1.25, 1.5, 1.75, and 2.0 at the levels of the four reflectors. Note that if they were dipping but parallel in the geological section, the seismic section would show their dips increasing in proportion to $\bar{V}$ from top to bottom of the section, in this case by a factor of nearly two.

(*d*) Depth in centre of section will be 2900 m, giving 100 m of relief, which is considerable in petroleum prospecting, showing the need for accurate velocity control.

A.3. The 2 m ms^{-1} layer presumably represents the outcropping shales, and the depths to the sandstone are $V_1 t_i/2 \cos \theta_c = 2 \times 75/2 \cos (\text{arc sin } 2/3) = \underline{101 \text{ m}}$ below *AB* and $2 \times 38/2 \cos (\text{arc sin } 2/3) = \underline{51 \text{ m}}$ below *CD*. The corresponding vertical two-way travel times in the shale are $2h/V_1 = 101$ ms and 51 ms respectively, leaving 199 ms in each case as the vertical two-way time in the sandstone. The sandstone thus appears to have a uniform thickness of $V_2 t/2 = 3 \times 199/2 = \underline{299 \text{ m}}$ and is faulted up by 50 m between *B* and *C*, the faulting being indicated by the curved diffraction events as well as by the up-step of

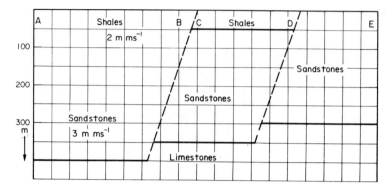

Fig. A.3

the reflector. Between D and E a fault of the same throw is seen to occur which will bring the sandstone to the surface. The refraction $t-x$ graph from D to E will therefore consist of a 3 m ms^{-1} branch preceded perhaps by a short branch at 2 m ms^{-i} depending on how the hade of the fault is extrapolated to the surface (see Fig. A.3). This extrapolation has been guided by the change from reflection to diffraction, which is difficult to localize exactly.

The refraction graph from E to D will again show mainly the 3 m ms^{-1} velocity, perhaps ending with a short 2 m ms^{-1} branch.

A.4. The first event on any marine reflection record represents the sea-floor, seen in Fig. Q.4 at a two-way time of 60 ms decreasing to 20 ms as the sandbank is crossed. The first branch of the refraction $t-x$ graph gives a velocity of 1.5 m ms^{-1}: that of water, as would be expected. Using $h = V_1 t/2$ we obtain water depths of 45 m and 15 m at the ends of the profile.

Event 2 is clearly the double reflection from the bottom, running from 120 to 40 ms, cross-cutting the deeper event, and always having twice the travel time and dip of the bottom primary. Once identified, it can be ignored.

Finally, we presumably have a reflection from the bedrock underlying the sand, and the concealed cliff-line is seen between A and B as a 10 ms step, with a diffraction event originating at its top. This diffraction shows that the "cliff" is too strongly convex or too steep to return energy as a conventional reflection, but does not indicate its true form. The beginning of the reflection from the base of the cliff, moreover, shows that this is no more than 10–20 m horizontally from the top, again indicating a steep feature.

The depths to bedrock can be found as follows: the two-way time in the sand layer is 45 ms below B, 55 ms below A, and 25 ms further to the left, giving sand thicknesses at these three positions of 45, 55, and 25 m using the sand velocity of 2 m ms^{-1} indicated by the second branch of the refraction $t-x$ graph. The depth to bedrock is obtained by adding the water depths of 15, 15, and 45 m, giving 60, 70, and 70 m for the depths below B, below A, and off the bank. The "cliff" is thus 10 m high, and the deepening of the bedrock to the left of A suggested by the form of the bedrock reflection is *not* real but is simply due to the increase in the thickness of the low-velocity water layer.

Construction of the refraction $t-x$ graph is straightforward, using equation (2.15) to find from the depths at A the two intercept times, which are 13 and 43 ms. The third branch, corresponding to the headwave from the bedrock, might be expected to show a downward step as the refractor depth decreases at the cliff, but this step is in fact half-concealed by the crossover with the second branch.

A.5. (*a*) The $t-x$ graph eastward from A begins with a 5 m ms^{-1} branch through the origin, corresponding to the direct path through the granite. About 50 m from A, a lower velocity of 3.6 m ms^{-1} is seen which must be the down-dip apparent velocity from the valley side: the corresponding up-dip velocity of 7.0 m ms^{-1} is seen as the second branch of the $t-x$ graph westward from B, which begins with a branch through the origin giving the drift velocity of 1.5 m ms^{-1}. West of the 50 m mark, the true granite velocity of 5 m ms^{-1} is again seen, confirming that this is the western edge of the valley. The slope of the valley floor is found using equations (2.16) and (2.17):

$$\sin\,(\theta_c + \alpha) = 1.5/3.6, \qquad \sin\,(\theta_c - \alpha) = 1.5/7.0$$

hence $\alpha = 6°$ and $\theta_c = 18\frac{1}{2}°$, giving $V_2 = 4.7$ for the true bedrock velocity. This is some 6 per cent lower than the velocity of the exposed granite, which is surprising but not in itself sufficient evidence for a bedrock change near the edge of the valley. The 6 ms intercept of the 7.0 m ms^{-1} branch at B implies a depth there of $6 \times 1.5/2$ cos $18\frac{1}{2}° = 4.7$ m. The fact that this depth is measured normal to the bedrock rather than vertically is not sufficient to account for the discrepancy with the borehole depth of 5 m, but again a 6 per cent discrepancy is within normally acceptable limits.

(*b*) In equation (3.7) we have $t_{ds} = 0$ (since bedrock is exposed at A) and so $t_{dg} = t - x/V_2 = t - x/4.7$, *assuming* that the lower V_2 found for the eastern half of AB persists across the valley. The delay times to the nearest millisecond are:

x m	50	100	150	200	250	300	350	400
$t_{dg} = t - x/4.7$ ms	− 1	2	4	8	11	11	− 1	5

If the velocity of 5 m ms^{-1} is used we obtain the following values:

$t_{dg} = t - x/5$ ms	0	3	6	11	15	15	5	0

These look rather more plausible than the first set, though the negative delay times are not significantly different from zero. None the less, the discrepancies of nearly 50 per cent between delays calculated using the two velocities illustrate the importance of *not* conducting a refraction survey in this way! A shot at the eastern edge of the valley to reverse the 400 m spread would have been vital.

The corresponding depths to bedrock are obtained *assuming* that the drift velocity is everywhere 1.5 m ms^{-1}. There is no firm evidence for this, and again one can see the deficiencies of the data: depths would require scaling approximately by the true local value of V_1.

x m	50	100	150	200	250	300	350	400
H m ($V_2 = 4.7$)	− 0.8	1.6	3.2	6.3	8.7	8.7	− 0.8	4.0
h m ($V_2 = 5$)	0	2.4	4.7	8.6	11.8	11.8	3.9	0

A.6. The curve of Fig. A.6.1 takes the reference point of the array to be the midpoint of P_1P_2, the usual convention. It is derived by the following argument.

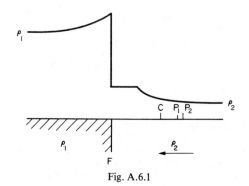

Fig. A.6.1

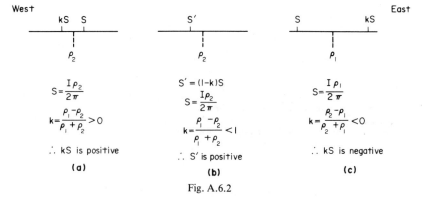

Fig. A.6.2

Consider Fig. A.6.2(a). Image theory assumes the boundary removed and the medium of resistivity ρ_2 replacing ρ_1. A positive image of the source, of strength kS, where $S = I\rho_2/2\pi$, is created as shown (see page 78).

As the array is traversed westward source and image converge, increasing the potential gradient across P_1P_2 and producing an increase in the apparent resistivity to $(1+k)\rho_2$ as the fault is approached. Once the source crosses the fault (Figs. Q.6(b) and A.6.2(b)), the potential array, still in medium 2, sees a source S' of constant but intermediate strength $S' = (1-k)S$. There are no images and there is no change in the apparent resistivity with distance from the fault. The apparent strength of the source increases discontinuously as the potential electrodes move into the same medium as the source, leading to a jump in the measured resistivity by a factor of ρ_1/ρ_2 (Fig. A.6.2(c)). A *negative* image is now "seen" by the potential array to the east of the fault, this producing an anomalously high potential gradient across P_1P_2 and thus an anomalously high apparent resistivity. As the array moves away from the fault so does the image, but in the opposite direction, its effect decreasing to zero, and the apparent resistivity dropping to the true value of ρ_1. The effect of any overburden would naturally be to smooth the sharper features of this curve.

A.7. The two-layer curve for the top two layers drops from 100 ohm-m to 30 ohm-m (Fig. A.7). Since $\rho_1 = 100$ and $\rho_2 = 33$ and $k = (\rho_2 - \rho_1)/(\rho_2 + \rho_1)$, then $k = -0.5$. As $h_1 = 10$ m, the exact curve can be drawn if curves are available. For a three-layer earth with a low-resistivity middle layer and a high-resistivity third layer current tends to flow horizontally in the top two layers and a replacement resistivity for these two layers can be found by treating them as resistances in parallel, the replacement thickness being equal to the sum of the thicknesses. The replacement resistivity is ρ_r and can be calculated from

$$\frac{H_r}{\rho_r} = \frac{h_1}{\rho_1} + \frac{h_2}{\rho_2} \quad \text{i.e.} \quad \frac{10 + 20}{\rho_r} = \frac{10}{100} + \frac{20}{33}$$

therefore $\rho_r \simeq 43$ ohm-m.

The lower part of the three-layer curve will approximate to a two-layer curve having the replacement layer ($\rho_r = 43$ ohm-m, $H_r = 30$ m) as top layer and a second layer of resistivity 300 ohm-m. For this layer $k \simeq 0.75$. The three-layer curve shown as a dashed line is accurately calculated for the given layering and it can be seen that it would be well matched by a visual smoothing between the two-layer curves.

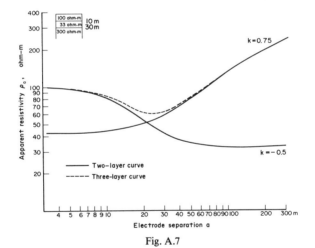

Fig. A.7

A.8. Soundings C and D are the same and indicate thick flat-lying strata with a resistivity of about 160 ohm-m below the dry sands. Nearby outcrop suggests these rocks are limestones, the curve indicating only a few metres (3 m) of overburden. To the west, at A and B the sediments are of lower resistivity (50 ohm-m) and at greater depth (10 m). The gravity profile indicates a steep contact between shales and limestones (assuming a contrast in rock densities of 0.5 t m^{-3} a fault with 100 m throw gives about the correct amplitude for the gravity anomaly). Changes in the thickness of the overburden suggest a small limestone scarp at the bedrock contact. A satisfactory electrode separation for the constant separation

AGG - P*

traverse would be about 30 m. This is large enough to clearly differentiate bedrock resistivities but not so large that the rise in resistivity as the fault is approached is so smooth that it is impossible to locate the change in rock type with any certainty. With the thicknesses of overburden present the profile will show at first a roughly constant value of 80 ohm-m, this rising smoothly as the fault is approached and levelling off over the limestone to a constant value of about 165 ohm-m. A small separation of, for example, 10 m would reflect merely changes in drift thicknesses since the measured values would be little affected by bedrock resistance.

A.9. Porosity is found from the expression $(1 - \phi)2.67 + (\phi \times 1) = 2.34$. For a unit volume $2.67(1 - \phi)$ is the weight of the sandstone matrix and $(\phi \times 1)$ is the weight of water in the pores, the total being the saturated density. $\phi = 0.2$, which gives a Formation Factor of 11.2. Since $F = \rho_{rock}/\rho_{water}$ and $\rho_{rock} = 77$ ohm-m, then $\rho_w = 6.9$ ohm-m. From the graph the NaCl concentration is 1000 ppm.

A.10. Density contrast is $2.4 - 2.2 = 0.2$ t m^{-3}. Range of the gravity anomaly is about 290 g.u. and the application of the "Bouguer slab" formula $\Delta g = 0.42\sigma \times h$ (from equation (7.9)) shows that the depth to the Triassic sandstones in the centre of the basin would be expected to be about $290/(0.42 \times 0.2) = 3450$ m. As the width of the basin is 20 km the assumption that it can be represented by a parallel-sided slab is not likely to cause such a gross discrepancy as that with the seismic data, which give a maximum thickness of Tertiary clays of less than 2000 m, occurring 4 km away from the gravity minimum on either side of it.

Using $\Delta g = 0.42\sigma + h$ again to calculate the gravity anomaly due to the known Tertiary thicknesses, we obtain the following table:

Distance (km)	0	2	4	6	8	10	12	14	16	18	20
Observed anomaly (g.u.)	35	80	157	240	280	290	280	240	157	80	35
Calculated anomaly (g.u.)	17	52	113	160	125	84	125	160	113	52	17
Discrepancy (g.u.)	−18	−28	−44	−80	−155	−206	−155	−80	−44	−28	−18

The gravity anomaly which remains to be accounted for is thus a symmetrical low centred on 10 km with an amplitude of about 200 g.u. The feature causing this low presumably lies within or below the Triassic: the presence of halokinetic features in the area suggests that a salt dome or ridge may be responsible. If this is correct, hydrocarbon prospects are enhanced since petroleum traps are often associated with such features. On the other hand, the low might be caused by a deep granite intrusion into the basement rocks, which would have a zero or negative effect on hydrocarbon prospects.

A.11. The range g_{max} of the anomaly is 15 g.u. and so the anomaly value corresponding to $g_{max}/2$ (i.e. to $g/g_{max} = 0.5$ on Fig. 7.8) is $5 + 15/2 = 12.5$ g.u. This is found at a point 540 m east of the origin, which on the simplest interpretation is the position of the fault trace. The sheet of Permo-Trias will have a thickness t given by $g_{max} = 0.42\sigma t$, i.e. $15 = 0.42 \times 0.250 t$, i.e. $t = 143$ m. As a check, supposing Coal Measures to be outcropping to the east of this fault, the Permo-Trias can be regarded as a "thin" sheet at a depth of $143/2 \simeq 72$ m. Remembering that the anomaly starts from a baseline of 5 g.u. it should therefore have values of $5 + g_{max}/4 = 8.75$ g.u. and $5 + 3g_{max}/4 = 16.25$ g.u. at distances of $z = 72$ m on either side of the fault trace. These "half-widths" are in fact about 160 and 145 m, confirming the evidence of the drilling that this interpretation is inadequate.

The fault at C can be approximated by a sheet for which $t = 238 \times 50 = 11.9$ t m^{-2} and $z = 25 + 50/2 = 50$ m. The given table can therefore be completed as follows, since $g_{max} = 0.42 \times 11.9 = 5.00$ g.u.:

x (m)	−200	−150	−100	−50	0	50	100	150	200
g (g.u.)	0.4	0.5	0.75	1.25	2.50	3.75	4.25	4.5	4.6

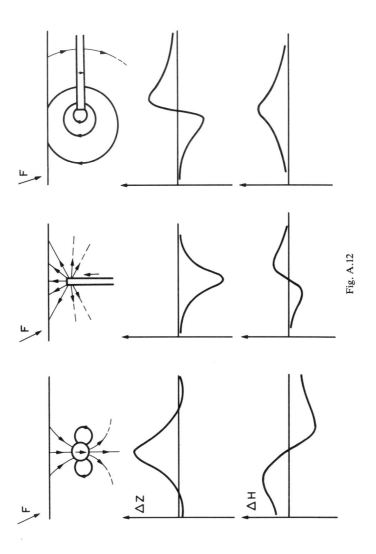

Fig. A.12

These anomaly values must be subtracted from the observed profile, placing the 0 m point at *C*, to allow for the now-known step in the unconformity. The remaining anomaly will of course have a range of 10 g.u., and is found to have a form closer to that expected for a simple fault. Repeating the interpretation procedure, we find a second fault at about 435 m from the origin, a throw of $10/(0.42 \times 0.238) = 100$ m, and a depth to centre of about 140 m from the two half-widths, in fair agreement with the expected $75 + 100/2 = 125$ m.

The presence of salt at *E,* to the west of this second fault, would increase the average density contrast between the two formations, so that the observed anomaly would in fact be produced by a *smaller* throw than 100 m.

A.12. Field lines for the three cases, and the anomalies deduced from them, are shown in Fig. A.12. Note that the sign of ΔH is taken as positive when the anomalous field is reinforcing the small horizontal component of the geomagnetic field, taken to be dipping steeply to the right. The vertical component anomaly is, of course, nearly the same as the total field component anomaly ΔF for the same bodies in this steep field, and the horizontal component anomaly is the total field anomaly that would be recorded if they were placed, with remanence unchanged, at the magnetic equator.

Author Index

Subject Index